Fundamentals in Organic Geochemistry

Series Editors
Jan Schwarzbauer
RWTH Aachen University, Aachen, Germany

Branimir Jovančićević
University of Belgrade, Belgrade, Serbia

Organic Geochemistry is a modern scientific subject characterized by a high transdisciplinarity and located at the edge of chemistry, environmental sciences, geology and biology. Therefore, there is a need for a flexible offer of appropriate academic teaching material (BSc and MSc level) addressed to the variety of students coming originally from different study disciplines. For such a flexible usage the textbook series 'Fundamentals in Organic Geochemistry' consists of different volumes with clear defined aspects and with manageable length. Students as well as lecturers will be able to choose those organic-geochemical topics that are relevant for their individual studies and programs. Hereby, it is the intention to introduce (i) clearly structured and comprehensible knowledge, (ii) process orientated learning and (iii) the complexity of natural geochemical systems. This textbook series covers different aspects of Organic Geochemistry comprising e.g. digenetic pathways from biomolecules to molecular fossils, the chemical characterization of fossil matter, organic geochemistry in environmental sciences, and applied analytical aspects.

More information about this series at http://www.springer.com/series/13411

Jan Schwarzbauer • Branimir Jovančićević

Introduction to Analytical Methods in Organic Geochemistry

Springer

Jan Schwarzbauer
Institute of Geology and Geochemistry
of Petroleum and Coal
RWTH Aachen University
Aachen, Germany

Branimir Jovančićević
Department of Chemistry
University of Belgrade
Belgrade, Serbia

ISSN 2199-8647 ISSN 2199-8655 (electronic)
Fundamentals in Organic Geochemistry
ISBN 978-3-030-38591-0 ISBN 978-3-030-38592-7 (eBook)
https://doi.org/10.1007/978-3-030-38592-7

This Springer imprint is published by the registered company Springer Nature Switzerland AG.
The registered company address is: Gewerbestrasse 11, 6330 Cham, Switzerland

Preface

The multidisciplinary field of organic geochemistry is clamped by its core tool, the organic analysis. The detection, identification and quantification of organic substances in the geosphere are a huge challenge. Organic compounds appear in complex mixtures at partly very low concentrations down to ultra-traces and cover a very wide range of properties, e.g. from lipophilic to polar substances or from low-molecular-weight compounds to macromolecules. Additionally, natural and xenobiotic organic substances in the geosphere are characterized by a high molecular diversity.

All these aspects require the complementary use of complex analytical methods. Therefore, organic-geochemical analysis uses a broad spectrum of techniques and approaches in varying systems. This volume as the fourth part of the book series *Fundamentals in Organic Geochemistry* aims at giving a brief but comprehensive introduction to these analytical techniques, the principal approaches of data interpretation and challenges and pitfalls organic geochemists are facing. The objective of this volume is not only to allow students to obtain basic and broad insights into analytical topics but also to start first steps in organic analysis.

Aachen, Germany — Jan Schwarzbauer
Belgrade, Serbia — Branimir Jovančićević

Contents

Chapter 1
Introductory Aspects

Organic Geochemistry as an interdisciplinary scientific field is methodologically based on organic analyses. Corresponding geoscientific problems are tried to be solved by analyzing organic substances in matrices from the geosphere, e.g. water, soil, rocks, air and to some extent also biota. Main intention is to obtain both qualitative and quantitative information on the composition of the organic matter. Or in general parlance: Which organic substance and how much of it appear in a certain geological or environmental sample?

Since individual organic substances occur not isolated but incorporated in organic and inorganic matrices, the analysis of such substances requires several separate steps as summarized in Fig. 1.1. It comprises roughly the sampling, the separation or even isolation of the target substance, the analyte, from the matrix, further cleaning and separation steps and finally the detection.

Main goal of any developed analytical method is the unambiguous identification of an analyte and the quantification with a sensitivity down to concentration levels, that are sufficient for the related scientific question.

The concentration level is partly a huge challenge for organic-geochemical analysis. This accounts especially for environmental contamination and its characterization. Concentrations determined in such samples covers a wide range as illustrated in Fig 1.2. In former times the so-called pp…-system has been used, where pp.. means 'part per …' ranging from ppm as 'part per million' down to ppq or 'part per quadrillion' and lower. However, this system is inaccurate since it is based originally on 'particle number per particle number' but has also used in dimensions like 'particles per volume' (e.g. mol/L), 'mass per volume' (e.g. μg/m^3) or 'mass per mass' (μg/kg). Therefore, the SI units have to be used instead of the old pp…-notation. Further on, concentration as mass per volume and amount as mass per mass have to be distinguished.

Beside the huge concentration range, organic substances cover a wide range of chemical and physical properties. And the same accounts for the geological and environmental matrices. As a consequence, there is no perfect analytical approach

J. Schwarzbauer, B. Jovančićević, *Introduction to Analytical Methods in Organic Geochemistry*, Fundamentals in Organic Geochemistry,
https://doi.org/10.1007/978-3-030-38592-7_1

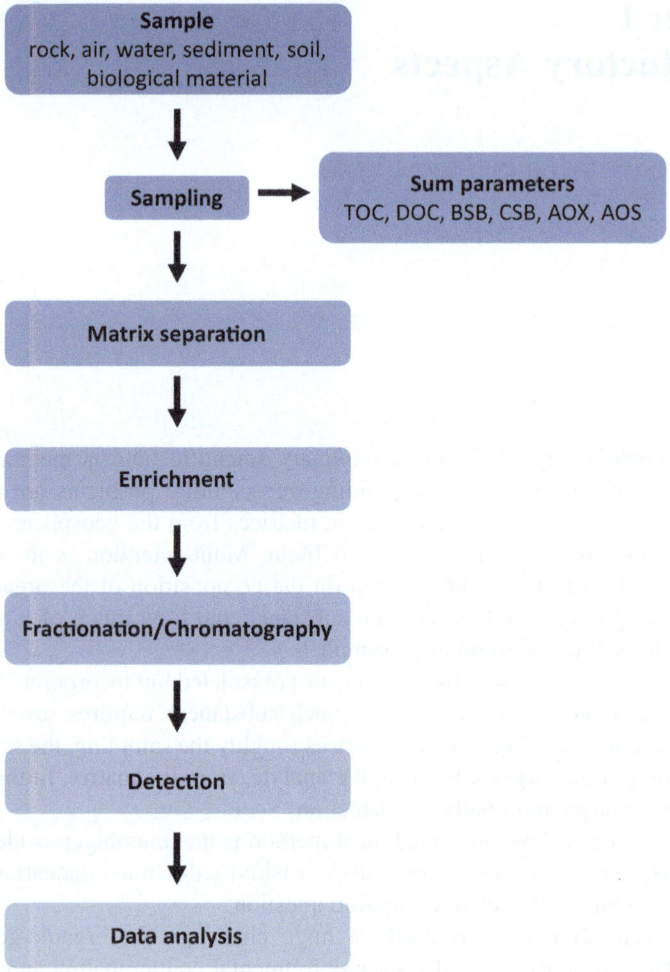

Fig. 1.1 Principal workflow of organic-geochemical analyses

applicable to all analytes in all sample types. Therefore, there is a need to become familiar with different analytical techniques and to become qualified to combine the individual techniques and approaches for a successful detection and determination according to the varying chemical properties of the numerous analytes. This is roughly illustrated in Fig. 1.3.

General Note

To analyse organic substances with high sensitivity and accuracy, a suitable analytical approach needs to be adopted to the chemical properties of the analytes. An universal method does not exist.

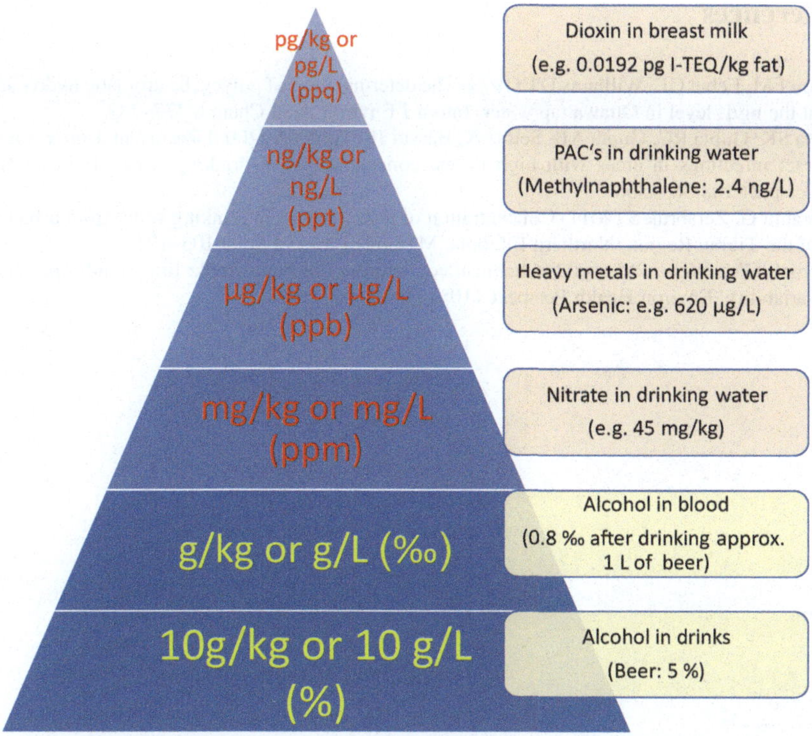

Fig. 1.2 Ranges of concentrations and amounts for organic analysis with some exemplifying values (data adapted from Benoit et al. 1979; Solomon and Weiss 2002; Gupta et al. 2000; Mebrahtu and Zerabruk 2011)

non-extractable → extractable

Substance properties	Chromatography	Detection
Medium to highly volatile lipophilic (non-polar to semi polar)	Gas Chromatography (GC) →	flame ionization detector (FID), mass spectrometry (MS)
	Liquid chromatography (LC) →	MS, infrared spectrometry (IR), ultraviolet–visible spectroscopy (UV-Vis)
Highly volatile, polar, macromolecular	LC (high performance LC, gel permeation chromatography)	UV-Vis, MS
Macromolecular	Pyrolysis-GC (or chem. degradation)	MS, FID

Fig. 1.3 Combination of analytical methods adapted to chemical properties of analytes

References

Benoit FM, Lebel GL, Williams DT (1979) The determination of polycyclic aromatic hydrocarbons at the ng/L level in Ottawa tap water. Intern J Environ Anal Chem 6:277–287

Gupta SK, Gupta RC, Gupta AB, Seth AK, Bassin JK, Gupta A (2000) Recurrent acute respiratory tract infections in areas with high nitrate concentrations in drinking water. Environ Health Perspect 108:363–366

Mebrahtu G, Zerabruk S (2011) Concentration of heavy metals in drinking water from urban areas of the Tigray Region, Northern Ethiopia. Momona Ethiop J Sci 3:105–121

Solomon GM, Weiss PM (2002) Chemical contaminants in breast milk: time trends and regional variability. Environ Health Perspect 110:A339–A347

Chapter 2
Sampling

Outlook
Sampling is an important task in all analytical procedures. The two main aspects that span the complex field of this issue, the sampling type and the sampling strategy, are discussed briefly in this chapter.

Sampling is one of the most fundamental parts in organic geochemical analyses. It is the base for reliable answers to the research questions that are initially intended to solve. Properly defined samples as well as an efficiently performed sampling are prerequisites for a reliable and successful entire analytical process—from the idea to the conclusions.

The complexity of sampling includes the type of material and corresponding sampling techniques as well as the strategy. The latter one covers e.g. frequency of sampling, number and quantity of samples. Since sampling is a complex issue a thorough introduction to this field requires a very comprehensive discussion. Further on, this book dominantly focuses on the laboratory work. Therefore, only a brief summary of sampling aspects is given in the following.

2.1 Sampling Type and Techniques

All samples studied in Organic Geochemistry can be roughly classified either by their origin (fossil matter, environmental samples) or their state of matter (solid, liquid, gaseous). Noteworthy, gas analyses play a minor role in Organic Geochemistry and are not handled here.

There are some clear differences between the individual sample classes, but also some overlaps. Differences are quite obvious e.g. between river water and river

J. Schwarzbauer, B. Jovančićević, *Introduction to Analytical Methods in Organic Geochemistry*, Fundamentals in Organic Geochemistry,
https://doi.org/10.1007/978-3-030-38592-7_2

sediment samples. These differences are dominantly the polar properties of the sample (water—polar, sediment—low to non-polar) characterized by hydrophilicity and lipophilicity, respectively, the concentration levels expected for the analytes therein and the homogeneity or heterogeneity of the matrix. Similarity in terms of sampling type can be observed for solid material. The sampling of a sedimentary archive in wetlands is highly similar to the techniques applied on coal sampling in open pit or underground mines. However, sampling needs to be adopted to the individual sample type.

Independently from the sampling type some preconditions need to be taken into account prior to any initial action. Any artificial contamination needs to be avoided. Main contamination sources are dirty sampling instruments or parts of them as well as already contaminated sample storage devices. Hence, it is important to preclean carefully all equipment used for sampling and sample storage. Further on, one major pitfall is to ignore immediate marking of the sample boxes for an unambiguous assigning of sample and sample location, layer etc. As a further aspect, the size or amount of a sample should be dimensioned to be as representative as possible and to contain enough material for performing all planned analytical procedures. Normally, a second sampling does not generate fully comparable material. Sample amount depends also on further specific properties of the material. As an example, if loose clastic sediments are sampled, then the sample has to contain enough matter within each particle size to adequately represent the size distribution. Consequently, samples of coarse material need to be substantially larger than fine sediments.

2.1.1 Fossil Matter

With respect to fossil material, oil sampling is a simpler task. Normally, these samples are taken on the oil fields, directly from the well. They often contain water and suspended solid material in addition to the oil phase. Sampling and further storage of oil samples are done in precleaned vessel (preferable glass ware). In order to prevent the loss of volatiles it is required to close the sample vessels immediately with a tight stopper. To take a representative sample, it is best to fill the bottle with crude oil in front of the nozzle at the oil well collector.

The second important sample type are rocks and sediments containing a certain amount of organic matter. Sampling procedure depends on the purpose of sampling, characteristic of sediments and location of sediment. Collected samples have to be undisturbed and representative of the whole rock or sedimentary system (e.g. spanning the whole grain size spectrum). It is also important to prevent any deterioration and contamination of the material during sampling.

One general sampling approach is coring. Several coring devices are in usage, that allows to obtain a more or less undisturbed stratigraphic column. For exploring geological archives covering long time periods on the level of hundred thousand to million years, the required deep penetration of down to kilometers below ground is achievable only by complex technical corers. Noteworthy, for many of such drilling

Fig. 2.1 An example of an outcrop area (**a**) and sampling work at an outcrop (**b**)

devices additives are used for an appropriate drilling. These drilling fluids pose the risk of sample contamination. This also accounts for the above-mentioned oil sampling from wells, that are also drilled with the support of drilling fluids.

An alternative sampling is related to outcrops, where stratigraphic systems are directly available (see Fig. 2.1). However, in this case the influence of weathering and oxidation processes need to be considered. Noteworthy, each sampling of sedimentary systems needs a thorough description of the lithology. Criteria of sub sampling sedimentary archives are either different lithological layer or regular intervals in case of homogeneous systems.

2.1.2 Environmental Samples

Principally, environmental organic-geochemical studies deal with water samples (surface and groundwater) as well as particulate matter such as soils, sediments and dust. Samples from surface water systems (rivers, lakes, sewage effluents ....) are taken by different often very simple scoop devices (see Fig. 2.2). Automatic sampling devices have been also developed allowing to sample e.g. an accumulated sample over 24 h by sampling a small aliquot at defined time intervals (e.g. each 10 mL each 10 min). Typical sample volumes are between 500 mL up to several liters depending on the concentrations of the target analytes.

Fig. 2.2 River water sampling by a telescope scoop device (**a**) and surface sediment catching at an estuarine area (**b**)

Fig. 2.3 Application and results of coring with a Geoslicer

Soil or sediment samples for environmental studies are either surface samples reflecting the recent state of pollution or archives used for reconstructing the pollution history. Surface samples are just grabbed by simple procedures, but for subaquatic sediments some technical devices have been developed, e.g. a Van Veen grab.

In principal the same technical applications as compared to fossil archives are used for environmental archives. Since the time period covered by environmental studies are much smaller (maximum of a few hundreds of years) the penetration depth is shorter with core lengths from centimeters to at least 100 m. Therefore, the drilling devices are often much simpler as compare to deep drilling techniques. Examples are a vibracorer or a geoslicer (see Fig. 2.3). However, also here it is an important precondition to obtain undisturbed cores or sample layers.

Fig. 2.4 Taking samples in a trench

The simplest way to obtain soil samples from archive layers is the preparation of a trench that is certainly limited e.g. up to 2 m depth. Here, grabbing a small trench produces a quasi-outcrop and allows an accurate and thorough sampling of individual layers after visual inspection (Fig. 2.4).

General Note
Sampling as a first but also a major activity in any analytical procedure, exhibits many pitfalls that need to be avoided. Further on, sampling needs to be designed according to the basic research questions the corresponding analyses wants to answer.

2.2 Sample Strategies

Taking one sample is already a sensitive task but taking a whole sample set in a proper and sufficient way is an intricate assignment. The main challenge is to fit the scientific research question with the sampling design in order to obtain at the end fully reliable and representative data. Since objectives of research studies vary to a high extend, there exist no general procedure or strategy for the design of a sampling campaign. Each campaign needs to be planned individually. There are some substantial parameters that need to be adapted: the sample size, representativity of sample locations, the sample grid or density of sampling locations and the sampling frequency.

However, based on a few examples some principal problems and the challenges behind are introduced in the following to illustrate the complicity of designing sampling campaigns. A first, very simple subject is the question about the amount. As already mentioned, there are some preconditions that the sample amount should fulfill, *inter alia* to be representative. one needs to decide about the sample size. Sample size can be also interpreted as aliquotation, since sampling only takes a part of the whole system investigated (e.g. a soil, sediments or coals). One main criterion for defining appropriate sample sizes or aliquotation rates is homogeneity. The more heterogeneous the matrix, the higher the variance by decreasing sample size. Or the higher the aliquotation rate, the lower the representativity in more heterogeneous systems. This relationship is illustrated in Fig. 2.5.

Sampling in extensive systems (e.g. coal seams, soil systems, coastal bay areas, river catchments, ...), especially with higher dynamics, requires an accurate decision on the position and the density of sampling locations. The following example will focus on the positions and partly on the density of sampling locations. If a river water section will be studied to trace the impact of a point source emission (e.g. effluents from a sewage treatment plant) one has to think about a suitable position of a reference site upstream of the emission point. Then the mixing of the contamination plume along the flow direction needs to be addressed, not only along the river course but also with water depth. A schematic suggestion for such research objective is given in Fig. 2.6.

A similar example but for a more static system is given in Table 2.1. If it is intended to trace a contamination in a soil system, the depth of sampling depends on the interaction and pathway of the pollutants that have to be studied. Only the top layers of a soil need to be sampled and analyzed, if erosion assisted distribution and

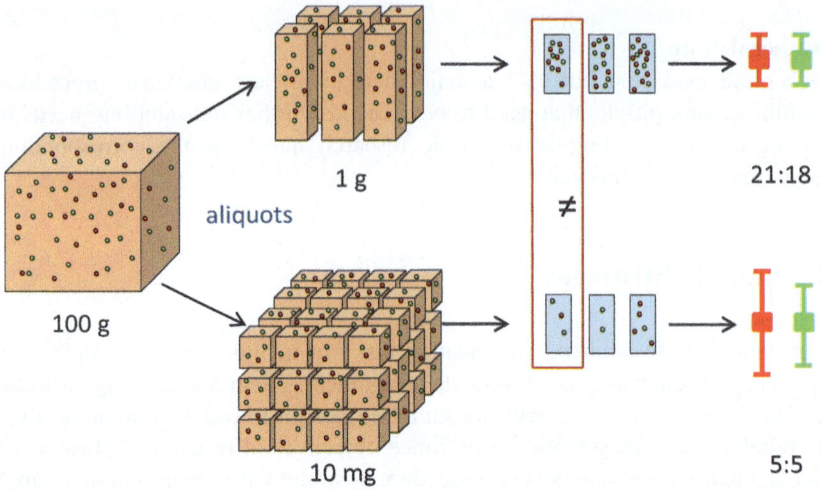

Fig. 2.5 Scheme exemplifying the relationship between aliquotation and representativity of samples

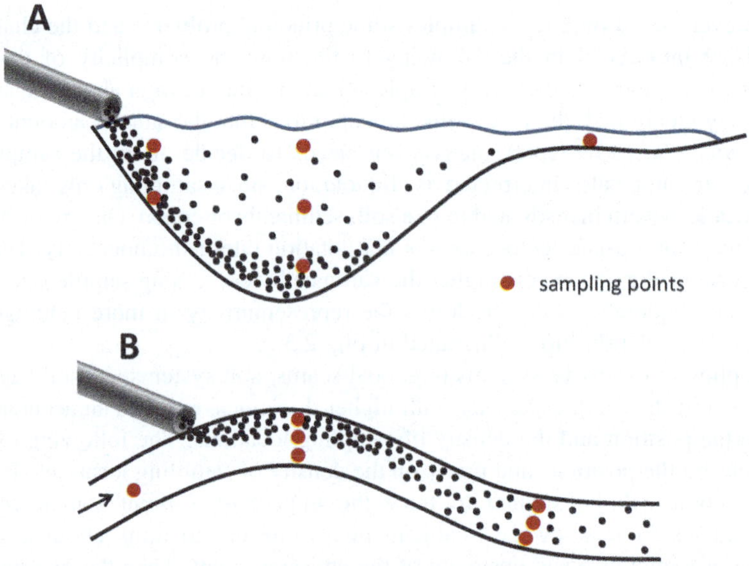

Fig. 2.6 Possible sampling strategy for analyzing river water affected by a point source

related processes will be investigated. For tracing the interaction of a soil pollutant with plant systems (e.g. plant uptake of pesticides), also the rhizosphere needs to be considered requiring sample depths down to 50–100 cm. Finally, if the potential for groundwater contamination by soil pollutants are of interested, the total unsaturated soil system down to aquifer has to be sampled.

Table 2.1 Pathway dependency of sampling depth in soils

Interface	Environmental pathway	Sampling depth (cm)
Soil–atmosphere	Erosion	0–10
Soil–human	Oral intake	0–35
Soil–plant	Plant uptake	0–70
Unsaturated zone–groundwater	Leakage	0–groundwater table

General Note

Sampling strategies and sampling design are huge tasks. Since they need to be closely adapted to the overall research question, there exists no general procedure or approach. A sampling campaign needs to be planned and developed for each study individually.

Chapter 3
Sample Treatment

3.1 Sample Storage and Pre-preparation

Outlook
Prior to the main analytical procedures two pre-routines need to be considered: sample storage and sample pre-preparations. Both aspects are discussed briefly here.

After sampling and prior to the main analytical process some preparation steps and storage of the samples normally require some attention (see Fig. 3.1). The order of the two procedures is flexible. The samples can either be stored prior to further treatment, or the treated sample material can be stored. However, the best practice is to subject the sample material immediately after sampling towards the initial analytical steps. However, in most cases this is not possible due to necessary long-distance transport, transfer to specialized laboratories or time constraints in the laboratory workflows.

Noteworthy, the raw sample material can also be subjected to various bulk analyses. Particulate matter can be analyzed e.g. for total organic carbon (TOC, see Sect. 4.4), Rock-Eval parameter (see also Sect. 4.4) or grain size distribution. For water samples e.g. the dissolved organic carbon and nitrogen (DOC, DON) or the AOX (adsorbable organic halogens) can be determined. However, in this textbook bulk analyses are not considered in more detail.

Typical pretreatment procedure for solid sample material are drying and grinding. Drying needs to be performed carefully in order to avoid loss of volatile components. Drying is performed (1) for an optimized extraction, (2) for a better storage since dried material is more resistant towards microbial attacks and, finally, (3) as preparation for the grinding process. Applying grinding improves the following extraction procedure due to an elevated surface of the particles that allows an enhanced

J. Schwarzbauer, B. Jovančićević, *Introduction to Analytical Methods in Organic Geochemistry*, Fundamentals in Organic Geochemistry,
https://doi.org/10.1007/978-3-030-38592-7_3

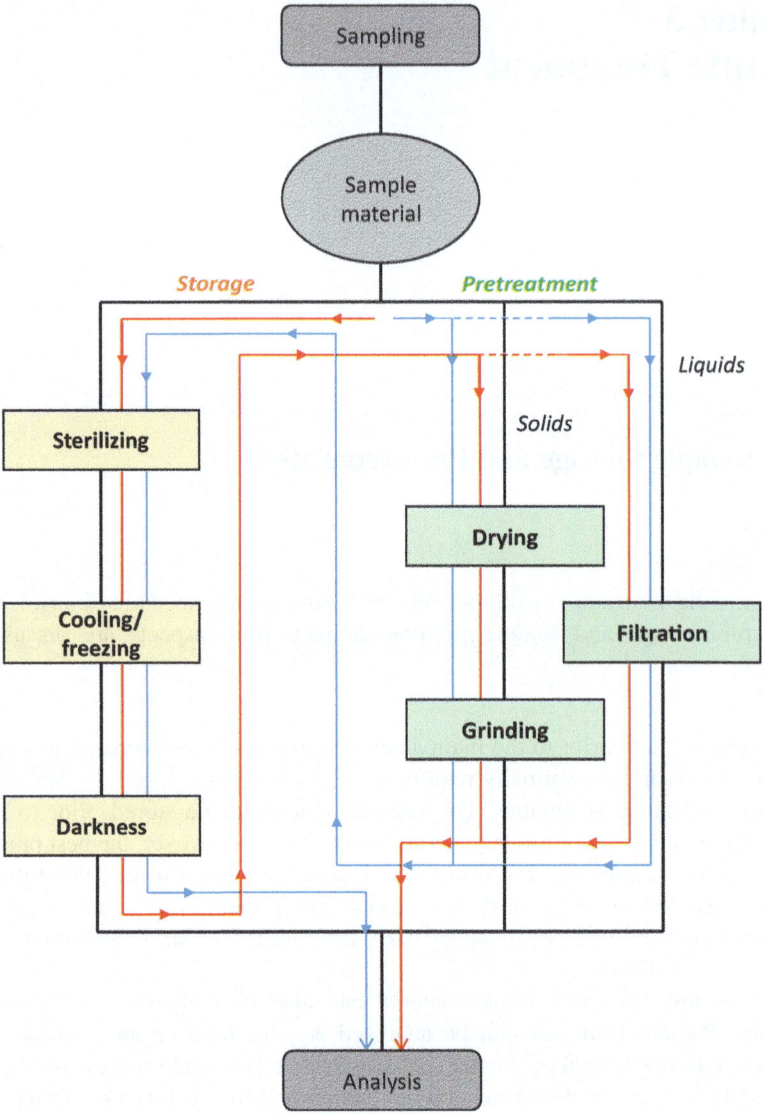

Fig. 3.1 Schematic workflow of sample handling after sampling and prior to the main analytical procedure. Note different pathways are possible (only pretreatment or storage (black), first storage followed by pretreatment (blue), or storage after pretreatment (red))

extraction yield (see Sect. 3.2). However, it has to be noted that information on the grain size distribution get lost during this process. Water samples are often filtrated prior to analyses. Commonly a filtration with 0.45 μm mesh size is performed, since this grain size defines the boundary between particulate matter and dissolved species.

Storage is a huge task, in particular if long-term storage is needed. There are different processes attacking the sample material over time comprising microbial transformation, heat and light induced changes as well as abiotic oxidation. Therefore, some protective measures need to be taken. In order to avoid alteration due to (sun)light and thermal stress, the samples should be stored in the dark and at low temperatures (e.g. 4°C or deep frozen). If anaerobic sample material needs to be stored (e.g. river sediments) anoxic conditions must be stabilized e.g. by addition of inert gas such as N_2.

The most critical point in storage of samples for organic analyses is to avoid microbial transformation of the matrix and, in particular, of the analytes. Biotic transformation is induced by both the microorganisms themselves and the corresponding extracellular enzymes. Hence both need to be reduced or eliminated, the microorganism (dominantly bacteria) and the enzyme activity. There are several methods for such a sterilization process that exhibit different sterilization strength, but they induce also changes of the sample material to a varying extent. The methods cover physical mechanisms such as γ-ray irradiation or autoclavation, but also chemical poisoning e.g. with mercury chloride $HgCl_2$ or sodium azide NaN_3. However, beside possible changes of the sample structure or integrity, also the duration for which the preservation is kept needs to be considered. Some examples of poisoning and the resulting maximum storage time for different target analytes in water samples are given in Table 3.1.

General Note
Storage and pre-preparation of sample material need some specific measures. Best practice is to perform the main analytical procedure immediately after sampling.

Table 3.1 Some examples of preservation and the resulting maximum storage time

Target analytes	Sample vessel	Preservation	Maximum storage time
Pesticides (organochlorine)	Glass	1 mL of a 10 mg/mL $HgCl_2$ or adding of extraction solvent (500 mL of water sample)	7 days, 40 days after extraction
Pesticides (organophosphorus)	Glass	1 mL of a 10 mg/mL $HgCl_2$ or adding of extraction solvent (500 mL of water sample)	14 days, 28 days after extraction
Phenolic compounds	Glass	Cool to 4°C add H_2SO_4 to pH < 2 (50 mL of water sample)	28 days

3.2 Extraction

Outlook
The first main step in the analytical procedure of organic geochemical analyses is the separation of organic analytes from the inorganic sample matrix. For this purpose, dominantly extraction methods are applied to solid and liquid samples using different organic solvents as extractants. The principals as well as the individual extraction techniques are introduced.

3.2.1 Principles of Extraction

The first important step in sample treatment is the isolation of the organic analytes from the sample matrix and a simultaneous enrichment. For these purposes the methods of extraction are the by far most common ones. These methods can be roughly classified by the state of matter of both the sample as well as the extracting agent or extractant (see Fig. 3.2). For sample material in organic geochemistry the solid and the liquid state are the most important ones. As extractants liquids as well as solid materials can be used. In combination, there exists one extraction system for solids (sediments, rocks, soils, . . .) that consists of solid sample material and liquid extraction solvents, shortly named *solid/liquid extraction*. For liquid sample material (mainly water) two extractions systems exist: the liquid/liquid and the liquid/solid extraction. Here, the application of solvents as extractants forming separate phases (liquid/liquid) or solid adsorption materials (liquid/solid) are used.

All extraction processes are based on a partition process between the two phases. After a certain time, this partition reaches a dynamic steady-state leading to constant

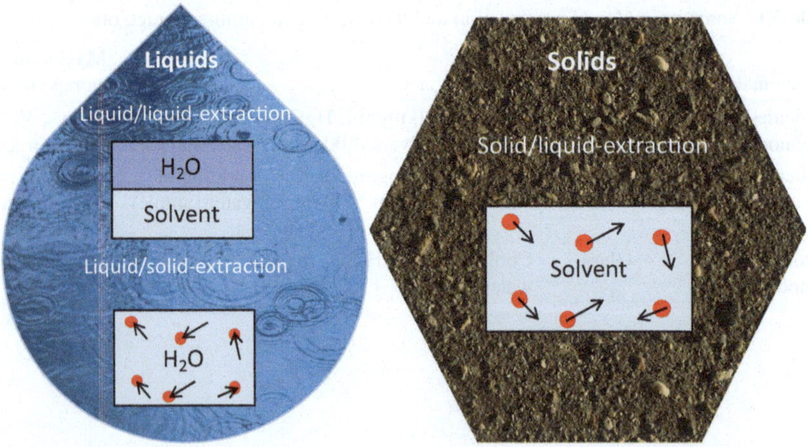

Fig. 3.2 Principal extraction systems

increasing polarity ↓

Compound formula	Group	Representative solvents	
R – H	Alkanes	Pentane, hexane	forming non-aqueous phase
Ar – H	Aromatics	Toluene, benzene	
R – O – R	Ethers	Diethyl ether	
R – X	Alkyl halides	Tetrachloromethane, chloroform	
R – COOR	Esters	Ethyl acetate	
R – CO – R	Aldehydes and ketones	Acetone	completely miscible with water
$R-NH_2$	Amines	Pyridine, triethylamine	
R – OH	Alcohols	Methanol, ethanol	
$R-COHN_2$	Amides	Dimethylformamide	
R – COOH	Carboxylic acids	Acetic acid	
H – OH	Water	Water	

Fig. 3.3 Commonly used solvents arranged by their polarity (comparable to the eluotropic series, see Sect. 3.3)

but different concentrations in both phases. Noteworthy, this steady state is specific for the both phases as well as the analyte. The constant ratio of both concentrations at equilibrium is expressed as partition or distribution coefficient that is specific for a certain system of phases and analyte. The coefficient is given in the following equation:

Distribution coefficient in steady state

$$K = \frac{C_{sample\ matrix}}{C_{extraction\ solvent}}$$

This approach implies several potentialities for optimization as well as restrictions. Due to the fact, that extraction works generally as a partition of analytes between phases under steady-state conditions, a 100% extraction yield cannot be achieved. A minimum concentration and an according amount of analytes remain in the sample matrix. However, the yield can be optimized generally by two different approaches. Firstly, an optimized selection of the extractant leads to a high partition potential for the selected analytes and, consequently, increases the extraction yield. As a rule of thumb, the polarity of the solvent should match well with the polarity of the analyte. In simple words, lipophilic compounds can be extracted best by non-polar extractants, whereas for hydrophilic compounds polar extractants should be used. For the usage of organic solvents as extractants, a list of commonly used solvents and solvent classes sorted by their polarity is given in Fig. 3.3. However, the selection optimizes the extraction process for one analyte, but might decrease the efficiency for another, chemically different one. Hence, in case of analyses of several target substances, an optimization for all analytes often needs a compromise in selection of the best extractant.

Secondly, the extracted amount can be enhanced by variation of the extractant volume, since only the concentration remains constant, but the total amount depends on concentration and amount. Increasing the volume of extractant has certainly some restrictions in technical handling and, in particular, the necessary reduction step of extractant volume after extraction. Further on, a one-step volume enlargement is not recommended, but a two- or multiple-step extraction with minor volumes (see example box).

Example Case: ***How to Dimension an Extraction for a Given Extraction Yield***

If you want to extract bis(2-ethylhexyl)phthalate (a plastizer widespread detected in surface water) from a one-liter sample of river water with hexane, you can estimate the amount of extractant you need for an extraction yield of more than 90 % based on the K_{OW} value:

$$K_{OW(DEHP)} = \frac{C\,(in\ extractant)}{C\,(in\ water)} = \frac{\frac{m}{V}\,(extractant)}{\frac{m}{V}\,(water)} \tag{3.1}$$

$$\log K_{OW(DEHP)} = 7.5 \quad \Rightarrow \quad K_{OW(DEHP)} = 0.875$$

Given data:

$m_{(extractant)} = 0.9$ (since a minimum of 90 % are requested to be extracted)
$m_{(water)} = 0.1$ (since only 10 % are left after sufficient extraction)
$V_{(water)} = 1000$ mL

Filling all data into the Eq. (3.1) and rearrangement to isolate the wanted term $V_{(extractant)}$ reveal:

$$V_{(extractant)} = 0.9 \cdot \frac{0.1}{Kow \cdot V(water)} = 0.9 \cdot \frac{0.1}{0.875 \cdot 1000\,mL} = 102.9\,\mathrm{mL}$$

$$\approx 100\,\mathrm{mL}$$

Hence, you need approx. 100 mL of hexane for getting an extraction yield of nearly 90% in one extraction step. Note, if you apply the 100 mL of extractant in two extraction steps (50 mL each), the extraction yield is higher. Check by yourself !

3.2.2 *Extraction Techniques*

As mentioned above, extraction methods are classified according to the state of matter of the sample material. For liquid/liquid extraction of liquid samples, in geosciences dominantly water samples, mainly two different techniques are used. A very common and simple approach is the direct extraction of water by shaking the sample with a defined amount of organic solvents acting as extractants in a separatory funnel as illustrated in Fig. 3.4. The mixture of sample and extractant needs to be rigorously and intensively shacked to disperse the solvent within the

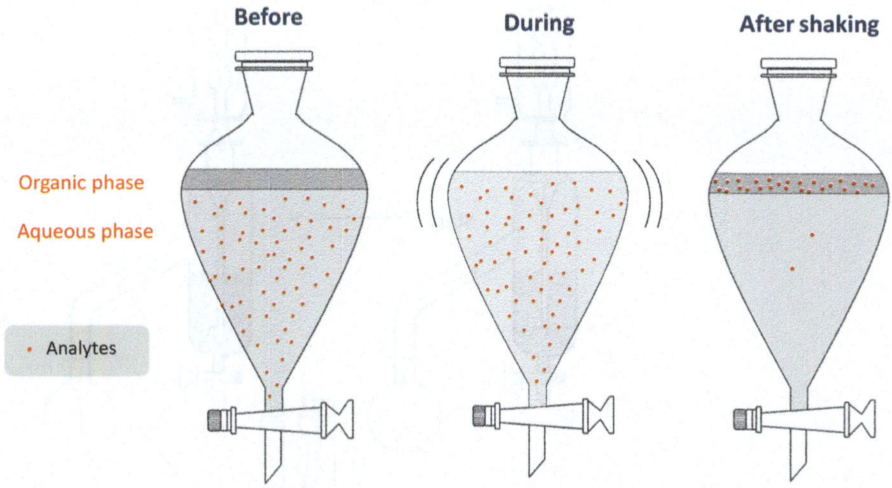

Fig. 3.4 A simple liquid/liquid extraction technique, shaking assisted extraction in a separatory funnel

water phase and to enhance the contact area between sample and solvent. A huge contact area improves the extraction speed and leads to a fast adjustment of steady-state conditions and sufficient extraction yield. After shaking the different phases can be easily separated and the organic layer can be collected. The application is suitable to run twice or in multiple steps. In this case, usage of different solvents as well as treatment of the water phase prior to extraction (e.g. change of pH, addition of salt) can extend the spectra of extracted substances enormously.

An obvious prerequisite for a successful extraction with this technique is the formation of separated phases, that means water and solvent shall not be mixable. In Fig. 3.3 a clear discrimination between miscible and non-miscible solvents is marked. This aspect restricts the selection of appropriate solvents, since more polar solvents (miscible with water) are not appropriate. As a consequence, ideal extractants for more polar substances (that means more polar solvents) are not suitable and the extraction yield will not be optimal.

Extraction by shaking exhibits the advantage of short extraction times (only a few minutes per extraction step) but the disadvantage of intensive personnel expenses and higher risk of lower reproducibility. A second technique is more automated, the perforator according to Kutscher and Steudel. A scheme is given in Fig. 3.5. Here a continuous extraction is achieved by dropwise penetration of the water sample with organic solvents. For this purpose, the extraction solvent is boiled in a receiver and the vapor is condensed into a system that allows to release very fine bubbles at the bottom of the water sample moving upwards. The solvent containing the extracted substances gets collected above the water sample and runs back to the receiver.

This technique can be run automatically but needs an elevated runtime of a couple of days (commonly between 48 and 96 h). Here, also the solvent can be changed during the extraction for extending the spectra of extracted analytes. However, one

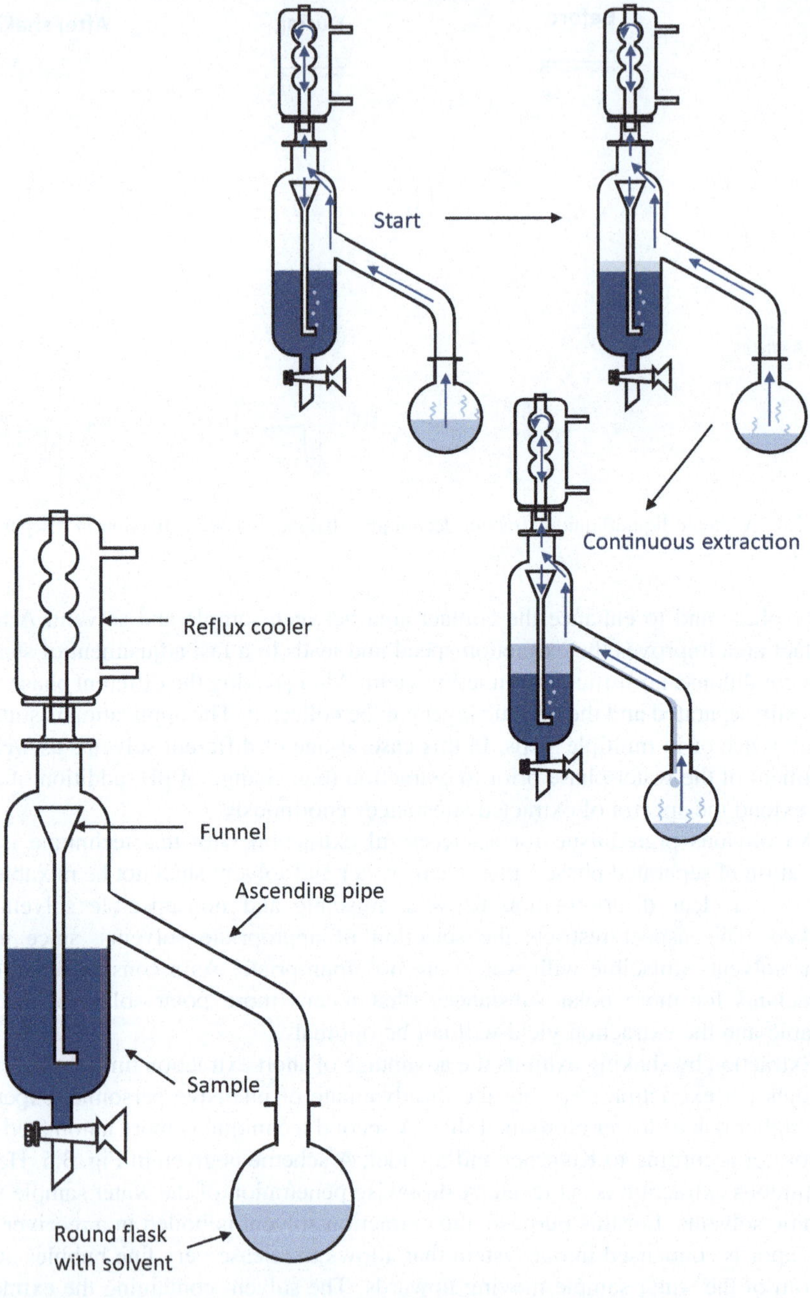

Fig. 3.5 Scheme of a Kutscher-Steudler perforator for a continuous liquid/liquid extraction

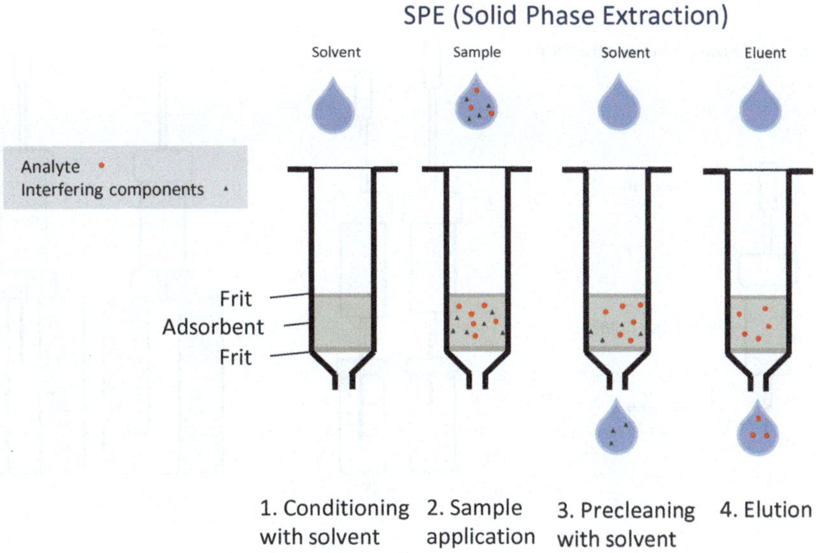

Fig. 3.6 Typical procedure of a SPE extraction

disadvantage is the continuous boiling of the solvent, in which the analytes are dissolved. This thermal stress avoids a successful enrichment of thermolabile substances. Further on, also for this technique the aforementioned restrictions regarding the selection of solvents are valid. Additionally, only extractants with a lower density as compared to water can be used, since they need to accumulate above the water sample surface.

As alternatives, various techniques of liquid/solid extraction techniques are used. Most common is the solid phase extraction (SPE) technique as illustrated schematically in Fig. 3.6. Here, solid adsorption material is used for the separation of organic analytes from water or related samples. The material is usually fitted in glass or plastic microcolumns (fixed with frits) and the liquid sample is passed through the solid material. The corresponding flow velocity as well as the chemical properties of the adsorption material are the most important parameters that need to be considered for a successful extraction. Adsorption materials cover a wide range of polarity from nonpolar reversed phases (e.g. C_{18} coated silica gel) to polar silica gels with various chemical modifications. Commonly, the SPE follows a four-step procedure starting with an activation or cleaning cycle, followed by the actual extraction step. Thereafter, the SPE columns need to be dried and finally, the adsorbed substances are discharged by elution with appropriate organic solvents. The last step can be modified by a sequential usage of solvents with different polarity (increasing or decreasing depending on the adsorption phase) for either a fractionation effect (for fractionation procedures see Sect. 3.3) or as precleaning step.

Since there is automatically a phase interface (solid–liquid) there are no restrictions for adsorption material polarity or elution solvents. Hence, a much wider range

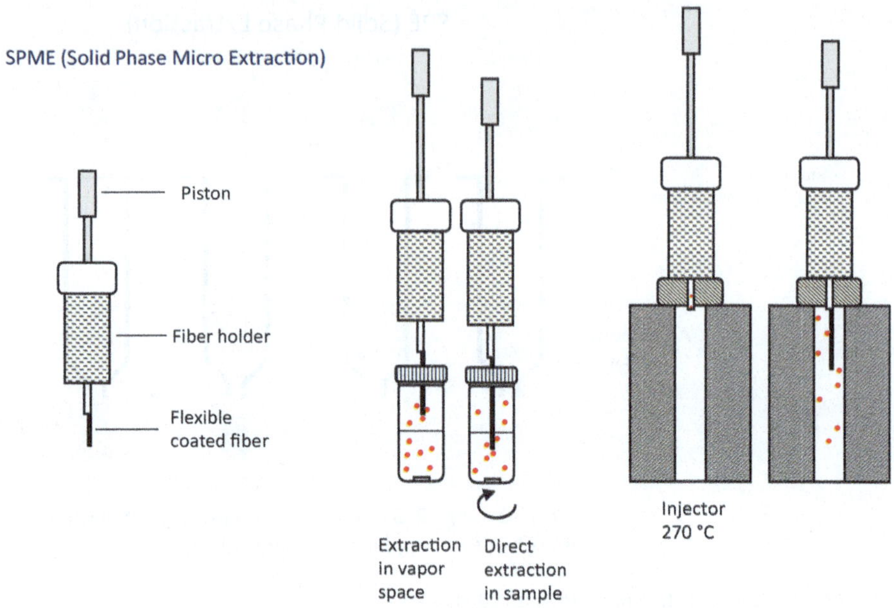

Fig. 3.7 Application process of SPME for liquid extraction

of substances in terms of polarity can be successfully extracted by this method. The only precondition is an insolubility of the adsorption material in both, water and solvents. However, disadvantages are a limited potential for automatization, a notable runtime of around 1–2 h and a more difficult and partly challenging selection of appropriate adsorption materials.

A variation of this method is the miniaturized technique solid-phase-micro-extraction (SPME). It is also a liquid/solid extraction that uses a very thin polymer fiber placed at the top of a syringe. The adsorption takes place directly in the sample by inserting the fiber into the stirred liquid phase. After a certain time of around 20–40 min, the syringe is lifted up and protects the fiber. Thereafter, the syringe needle including the fiber can be directly injected into a gas chromatograph (GC) and due to the high temperature in the injector (around 250–300°C) the adsorbed analytes can be directly vaporized and transferred to the GC column for further analyses (see also Sect. 4.1.1 for further information about GC). According to this procedure, the technique has been developed for volatile to semi-volatile substances and for a direct transfer to a gas chromatograph. Therefore, polar or low-volatile substances are excluded and more complex matrices, that need further fractionation or cleaning steps prior to GC measurements, are not suitable for SPME. However, the method is fast, needs no further solvents and avoids external contaminations.

Beside the direct SPME of the liquid phase, also a head space approach can be used. Here, the vapor space above the slightly heated liquid sample is extracted by the fiber and, consequently, only the high volatile analytes that accumulate in the head space, are enriched on the fiber for further analyses (Fig. 3.7).

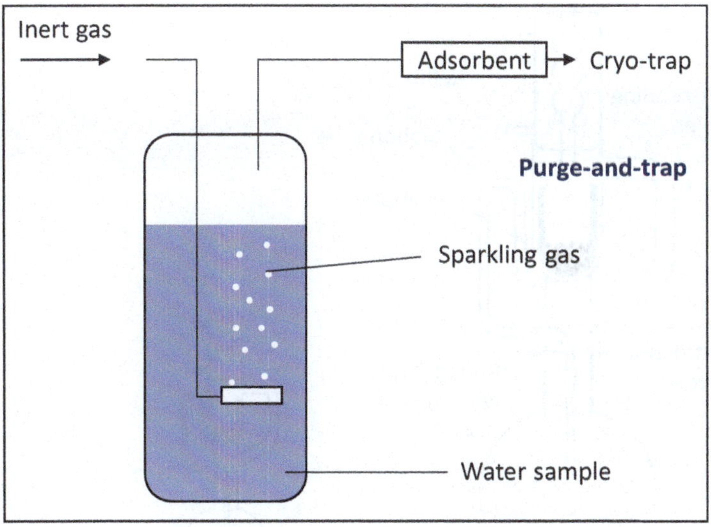

Fig. 3.8 Principal scheme of a purge-and-trap device for extracting high volatile components from liquid samples

A more special technique for extracting volatile substances from water samples is the purge-and-trap technique (see Fig. 3.8) which focuses on the same substance spectrum as compared to the head-space SPME. The volatile substances are purged by bubbling of an inert gas from bottom to top. During this process the volatile compounds are purged out into the gas phase and, in a second step, these components become adsorbed on a solid phase or become entrapped in a cyro-trap.

Similar to the extraction techniques applicable to liquid samples, the enrichment of organic analytes from solid matrices can be performed by various methods. However, all these methods belong to one system, the solid/liquid extraction. Since solids and liquids are *per se* different phases, less attention has to be paid to the extractants in terms of miscibility. However, comparable to the liquid/solid extraction systems, the only prerequisite is the insolubility of the solids in the extractants. As one important aspect, more or less all extraction methods for solids need fine grained sample material for a sufficient and reproducible extraction yield.

In Organic Geochemistry of fossil material one method has been used for decades, the Soxhlet extraction (see Fig. 3.9). This approach has a high similarity to the perforator described above. The solvent used for extraction is boiled and the vapor is condensed directly above the sample, that is filled into a semipermeable tube. This tube allows a penetration by liquids but is impermeable for particles. The condensed solvent drops into the sample, starting the extraction process. In parallel, the tube holder starts to get filled with the solvent and the extracted components. After reaching an upper level defined by the bend of an attached small glass tube. After reaching this upper limit, the total solvent in the tube holder gets drained into the receiver. Then a next extraction cycle starts.

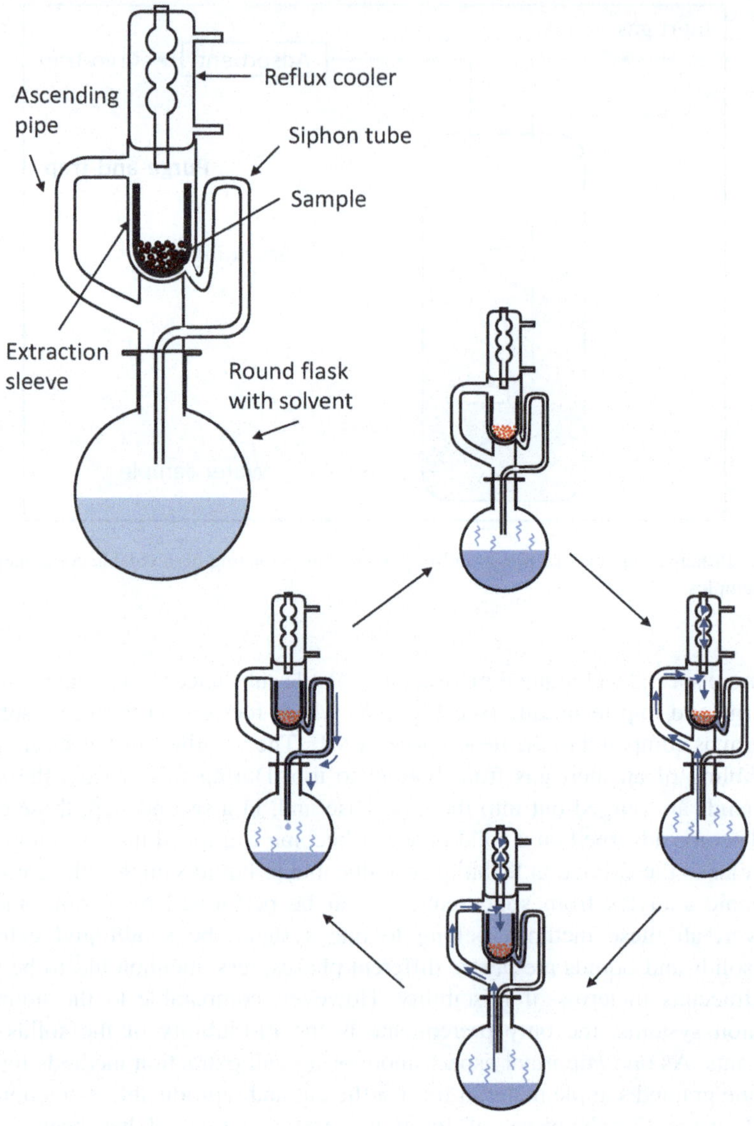

Fig. 3.9 Scheme of a Soxhlet device for the continuous extraction of solid pulverized samples

Advantages and disadvantages of this method are similar to the perforator system: Long extraction times (48–96 h), suitability only for thermostable compounds *versus* automatization and relatively high reproducibility. Noteworthy, for fossil matter the thermal stress do not need to be considered since fossil material is intrinsically thermostable.

A by far faster technique has been developed in the 1990s, the accelerated solvent extraction ASE. The sample material is extracted under high pressure and with

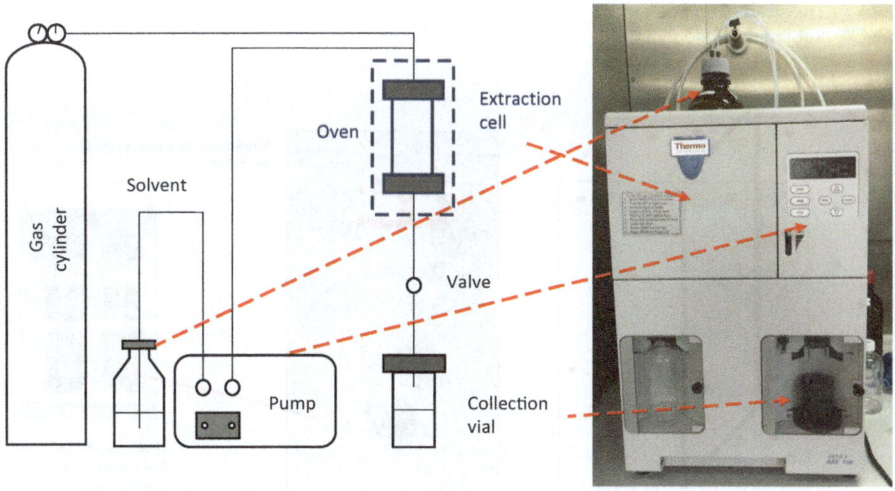

Fig. 3.10 The principal composition of ASE device

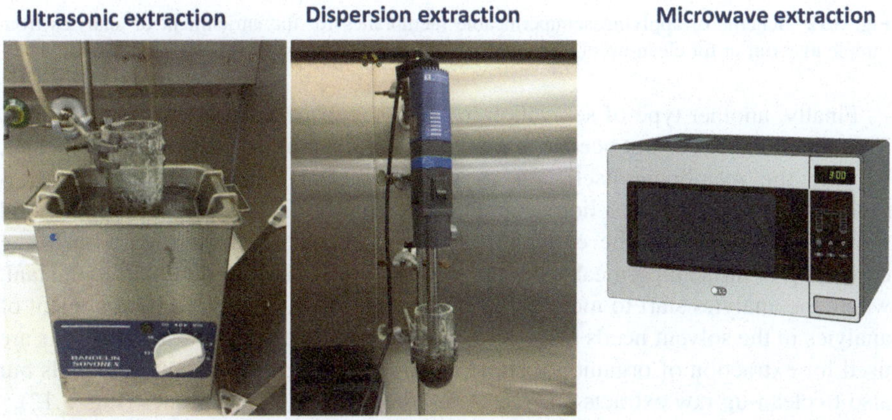

Fig. 3.11 Various types of extraction approaches using external energy supply for improving the extraction process

elevated temperatures for a short time period of only a couple of minutes (see Fig. 3.10). Temperature as well as pressure speed up the extraction leading to a sufficient yield in short time and can be varied adapted to the extraction problem. Further advantages are the small amount of extractants (typically 5–40 mL) as well as the possibility to fully automatize the procedure (as illustrated in Fig. 3.10).

However, there are further extraction techniques also using the external input of energy to accelerate and optimize the extraction. These techniques just differ in the type of energy comprising ultrasonication, microwaves or stirring energy. The corresponding implementations are illustrated in Fig. 3.11.

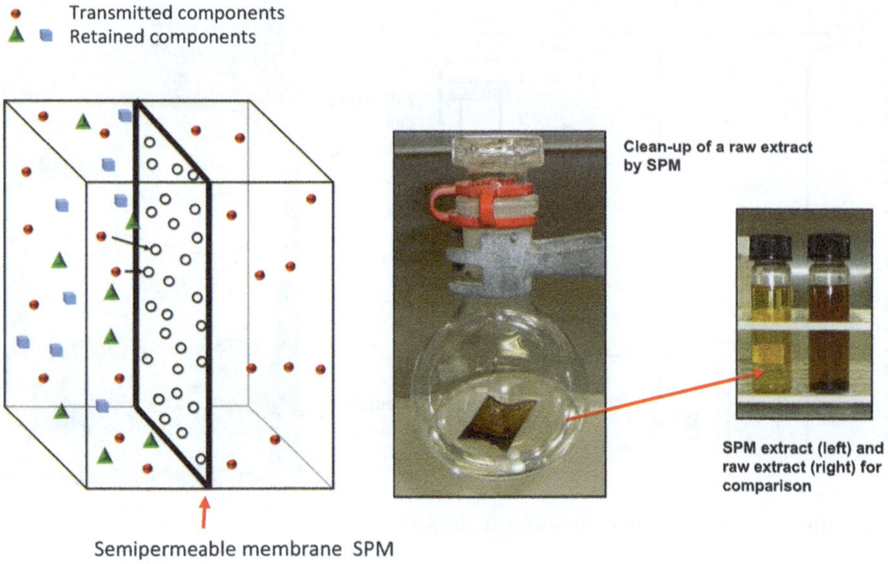

Fig. 3.12 Scheme of applying semipermeable membranes for the enrichment of analytes from sample material or for clean-up of raw extracts

Finally, another type of separation is using semipermeable membranes (SPMs). This application does not need a phase separation of sample material and extractant, because the membrane itself acts as separator. The enrichment of analytes is typically performed with a liquid sample or a slightly dissolved or moistened solid sample. For executing the extraction process, the material is enclosed in a bag consisting of the semipermeable membrane and this bag is placed into the extractant, where the analytes start to move through the membrane. A sufficient enrichment of analytes in the solvent needs some time, typically minutes to a few hours. SPMs are used for extraction of organic substances from food, biota, sediments and soils but also to clean-up raw extracts derived from other extraction techniques (Fig. 3.12).

General Note

Extraction method are optimized either by an enlarged extraction time that allows to reach the partition equilibrium or by fastening the dynamic partition process via external energy supply. Further on, the extraction process can be accelerated by a high contact surface e.g. by dispersing extractants forming very small droplets with overall high surfaces.

3.3 Fractionation

Outlook
The second step of organic-geochemical analyses is the coarse fractionation of the raw extracts. This process is commonly based on a chromatographic separation. Principally, gas and liquid chromatography are used in Organic Geochemistry.

3.3.1 Principles of Chromatography

As a result of extraction procedures, one obtains a complex mixture of organic substances in a defined organic solvent. For further analytical measurements, these mixtures are normally too complex to be analyzed directly. Therefore, a second main sample treatment step is needed, the fractionation of the raw extract into several subfractions. Basic method for the separation is the chromatography.

Chromatography works principally with two phases, the stationary and the mobile phase (see Fig. 3.13). The latter one flows through or passes alongside the stationary phase. The mixture of analytes is commonly dissolved in the mobile phase and is injected at the starting point of the system. Then the analytes get transported by convection with the mobile phase through the chromatographic system. Due to this process, the mobile phase is also called eluent. The convection is superimposed by dispersion of the analytes as the result diffusion of the individual molecules within the mobile phase. The extent of dispersion depends on the flow time and forms the typical shape of chromatographic signal, the peak. In average some of the molecules have a summarized diffusion pathway with a direction parallel to mobile phase flow resulting in a slight faster movement and forming the back tailing of the peak. Molecules with a summarized pathway smaller than the average flow effect the front tailing. As a result, chromatographic peaks look ideally like a symmetric normal distribution. Finally, the most important effect of chromatography called retention provokes the actual separation of different substances. The retention is based on an interaction of the analytes with the stationary phase. These interactions can vary widely and comprise e.g. ionic interaction, van-der-Waals forces or partition processes. If the analytes interact to a different extent with the stationary phase, the analytes become retarded divergently and get separated. It is obvious, that the interaction depends on the chemical and physico-chemical properties of the analytes and both chromatographic phases.

Consequently, the selection of the chemical composition of stationary and mobile phase is an important parameter to optimize the separation efficiency. All chromatographic techniques can be divided roughly by the state of matter of both phases (see Fig. 3.14). Dominant classification parameter is the mobile phase defining gas and liquid chromatography. A more special technique uses supercritical fluids as eluents

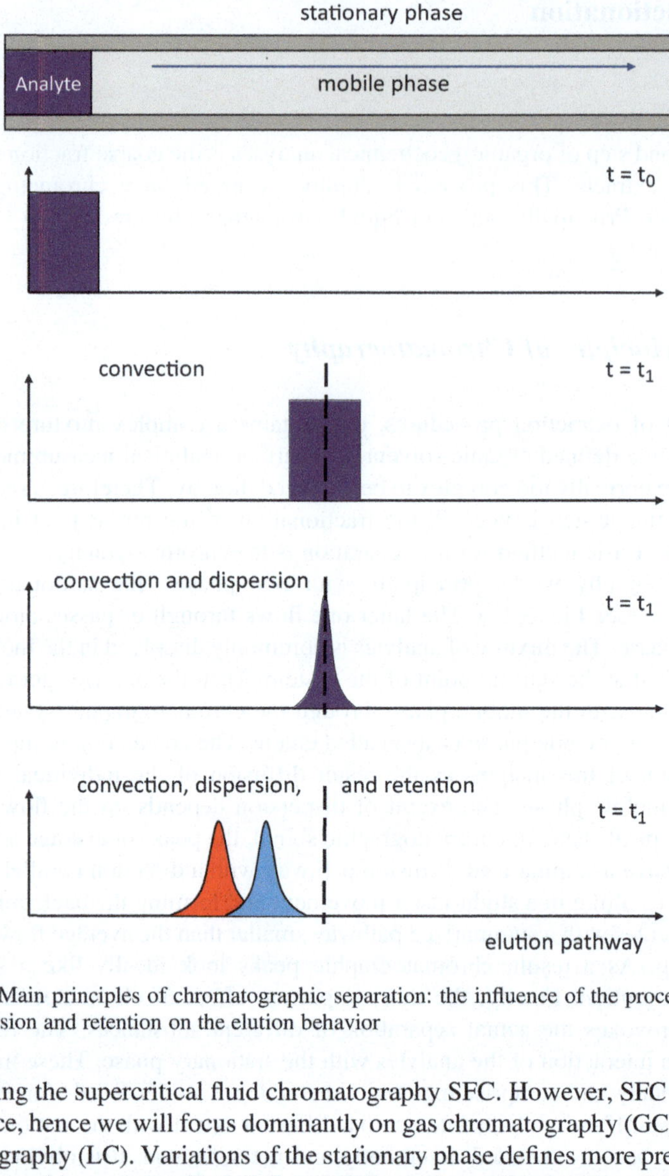

Fig. 3.13 Main principles of chromatographic separation: the influence of the processes convection, dispersion and retention on the elution behavior

representing the supercritical fluid chromatography SFC. However, SFC is of minor importance, hence we will focus dominantly on gas chromatography (GC) and liquid chromatography (LC). Variations of the stationary phase defines more preciously the chromatographic systems. Both LC and GC can be realized with solid and liquid stationary phases, whereby a special preparation of the liquid stationary phase is necessary for immobilization. Finally, the type of interaction between analyte and stationary phase discriminates the individual chromatographic techniques, e.g. ion exchange or size-exclusion chromatography.

Noteworthy, the different individual chromatographic techniques use various specific stationary phases and a wider spectrum especially of liquid mobile phases. This will be introduced in more detail in Sect. 4.1.

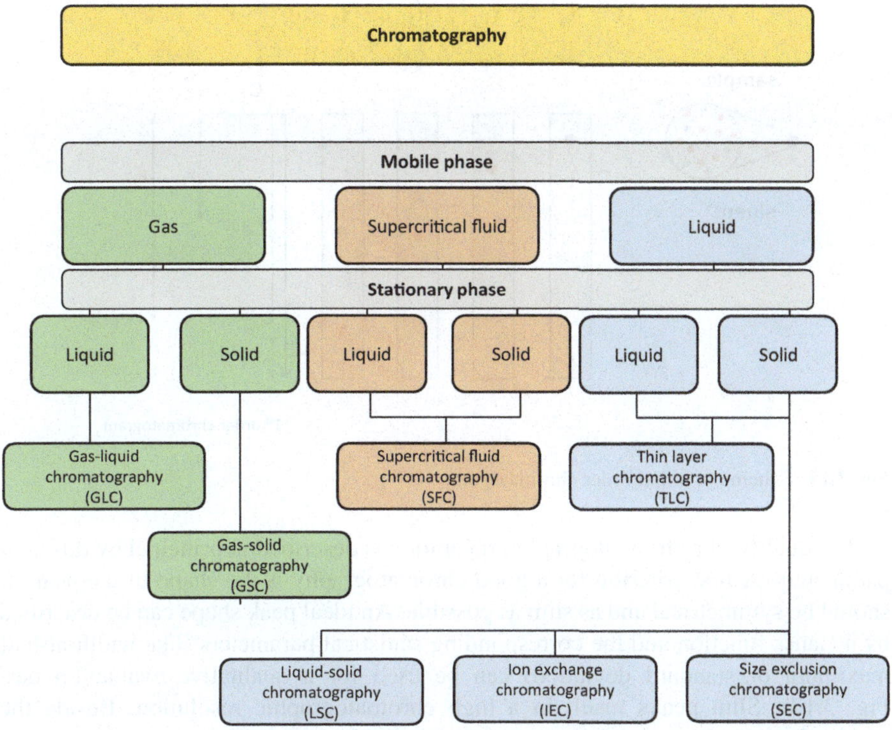

Fig. 3.14 Simplified classification of chromatographic systems (modified after Cammann 2010)

A further aspect characterizing the variability of chromatographic applications is related to the type of chromatograms. Chromatograms represent the result of a chromatographic separation. However, the separation can be obtained in two different dimensions, a spatial and a temporal one. First order chromatograms are a function of concentration and retention distance. Such chromatograms result from chromatographic separations that are stopped after a certain time as illustrated in Fig. 3.15. Then, the individual analytes are located at different distances from the starting point, the corresponding characteristic parameter as used in chromatography is the retention factor R_f. This value is the ratio between the actual retention pathway of an analyte and the maximum possible pathway. Correspondingly, these values range between 0 and 1 and represent the relative movement of an analyte in the given chromatographic system.

In a second order chromatography the analytes are allowed to leave the chromatographic systems. The time of elution, also called retention time, is measured and correlated with the corresponding concentrations. Hence, a 2nd order chromatogram is a function of concentration and time (see Fig. 3.16). An advantage of such chromatography is the possibility to collect individual fractions discretely. Thereafter, the separated compounds can individually be analyzed (see Sect. 3.3.2).

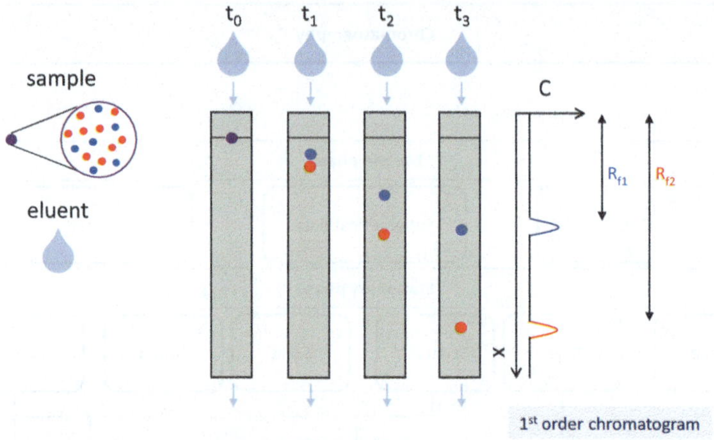

Fig. 3.15 Scheme of a first order chromatogram

The quality of a chromatographic separation is described in principal by different parameters. A first criterion for a good chromatography is the shape of the peak. It should be symmetrical and as slim as possible. An ideal peak shape can be described by a Gauss function and the corresponding statistical parameters (like width-at-half maximum or standard deviation) can be used for a qualitative evaluation (see Fig. 3.17). Slim peaks result in a high chromatographic resolution. Beside the width of a peak the separation 'distance' of individual peaks is the second important criterion, often called separation capacity. Both chromatographic properties, resolution and separation capacity, can be influenced by various individual parameters of a distinct chromatographic technique. The general influence of both factors is illustrated in Fig. 3.18. However, both parameters interfere, since a good separation capacity often results in longer chromatographic times, which induces broader peaks (lower resolution) due to the increase of diffusion processes with time. Hence, optimizing the efficiency of a chromatographic separation needs a very balanced improvement of both principal parameters.

General Note
Applying successfully chromatography needs a principal understanding of the main factors influencing the chromatographic efficiency. The resolution and the separation capacity.

Chromatography is used in analytical approaches in two different modes of application. High-performance chromatography such as gas chromatography (GC) and high-performance liquid chromatography (HPLC) are used directly linked to powerful detector systems such as mass spectrometry (MS). These applications are also known as hyphenated techniques and are discussed in detail in Chap. 4. Low-performance chromatography is often used for a coarse separation of raw

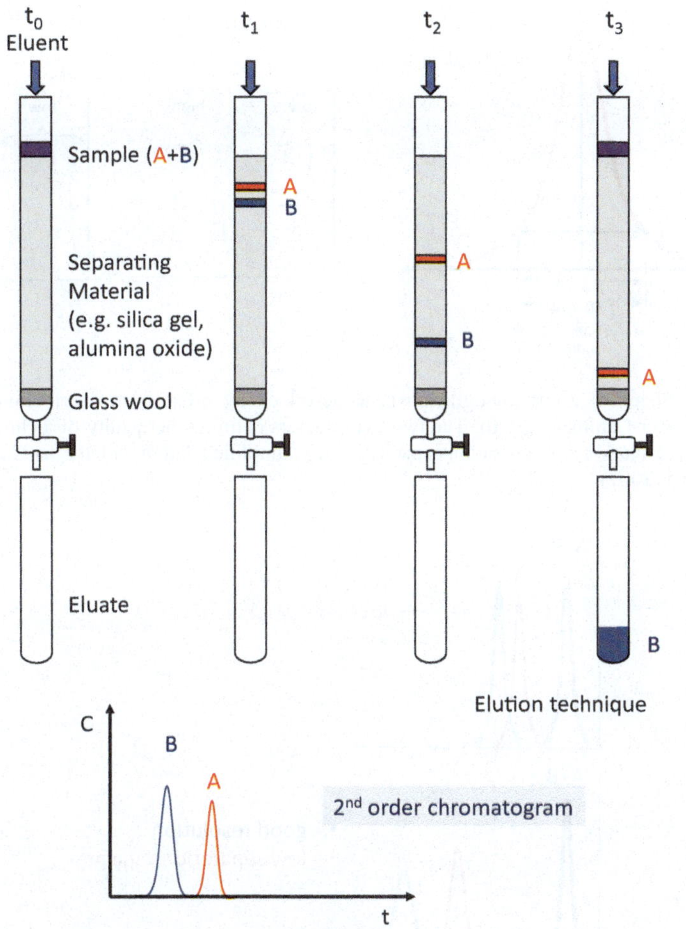

Fig. 3.16 Scheme of a second order chromatogram

extracts, subdividing the whole spectra of extracted substances into fractions according to the separation mechanism used. This approach is called fractionation and is introduced in the following.

3.3.2 Fractionation by Low-Performance Liquid Chromatography

A prerequisite for any successful chromatographic fractionation is a preconcentration of the raw extracts. Depending on the extraction method the raw extracts can have a few milliliters up to 500 mL. These volumes are commonly

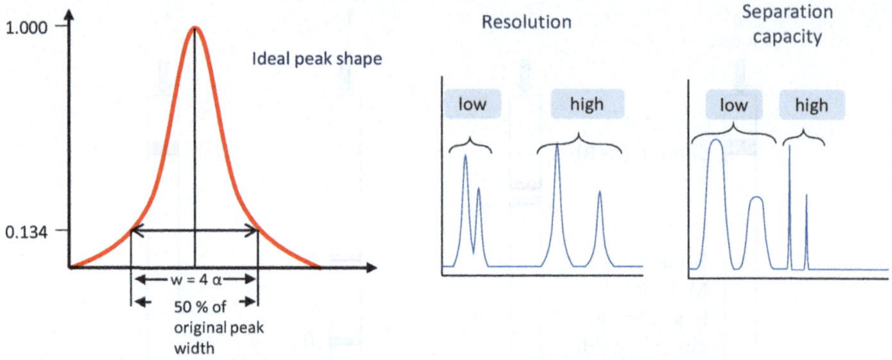

Fig. 3.17 Shape of a chromatographic peak described by a Gaussian distribution curve (left, simplified after Cammann 2010). The two main factors defining the quality of a chromatographic separation, resolution and separation capacity (right) (simplified and modified after Cammann 2010 and Schwedt 2007)

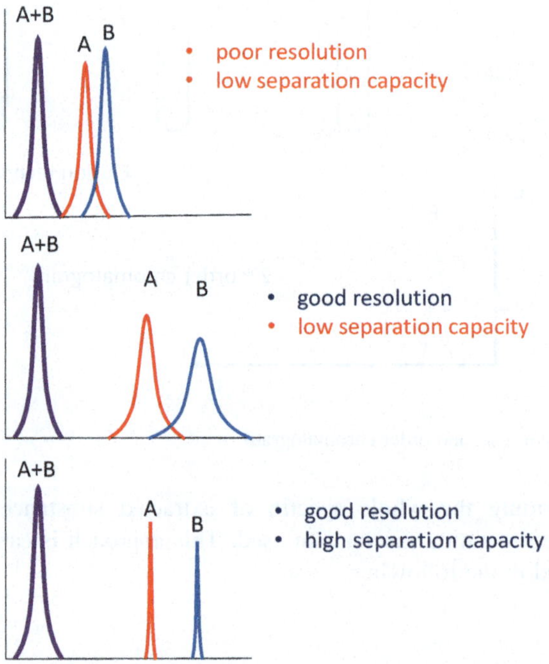

Fig. 3.18 How resolution and separation capacity influence the efficiency of a chromatographic separation (adopted from and modified after Schwedt 2007)

reduced to volumes of approx. 0.5–2 mL by rotary evaporation prior to fractionation or another coarse separation.

For fractionation the application of first order chromatography is limited. Thin layer chromatography is the most common practice using sheets (glass, plastic or

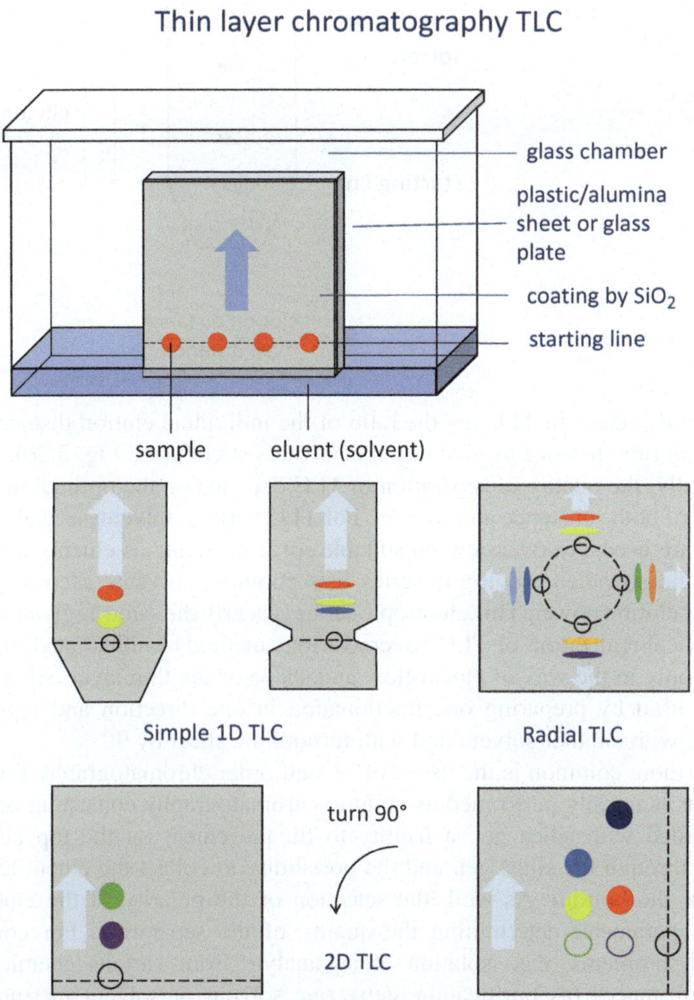

Fig. 3.19 Principles and various technical realizations of thin layer chromatography (adopted from and modified after Schwedt 2007)

alumina) covered with thin layers of silica gel (see Fig. 3.19). The separation is performed by stippling the compound mixture near the bottom of the sheet and by positioning the sheet into a solvent in a way that the lower end is placed in the solvent (without any contact of the substance mixture). Caused by capillary forces the solvent starts to move upwards and serves in this manner as the mobile phase. The individual substances become mobilized with different velocity depending on the polarity of the solvent and their own polarity. Since silica gel represents a polar phase, lipophilic compounds get mobilized relatively fast by nonpolar solvents, whereas the more polar substances need more polar mobile phases for mobilization. As already introduced, individual R_f values can be determined for each separated

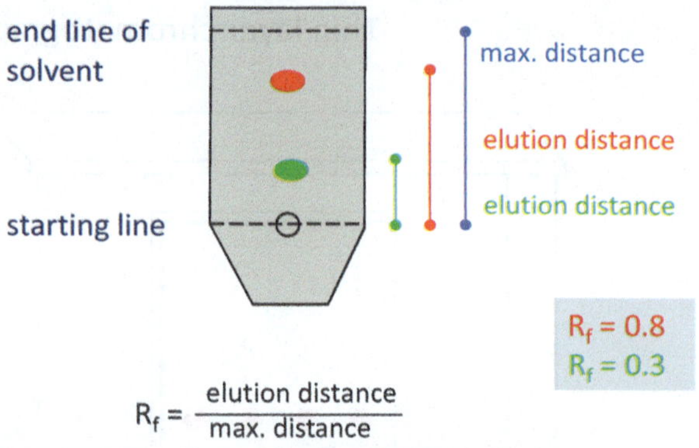

Fig. 3.20 How to calculate R_f-values in TLC

substance. R_f values in TLC are the ratio of the individual elution distance and the maximal elution distance as given by the solvents stretch (see Fig. 3.20).

Generally, the quality of separation in TLC depends on the optimal matching of polarities of both substance and solvent. For TLC various solvents as well as solvent mixtures are used. An overview on suitable solvents acting as eluents with various polarity, the so-called eluotropic series lists common solvents according to their polarity or elution power. This eluotropic series is nearly the same as given in Fig. 3.3. The technical realization of TLC covers various methods as illustrated in Fig. 3.19 differing only in the way of eluent flow and shape of the thin layer. Also, 2D-TLC has been used by preparing one fractionation in one direction and repeating the procedure with another solvent and with turning the sheet by 90°.

By far more common is the usage of second order chromatography for fractionation. This is usually performed as column chromatography consisting of a (glass) column filled with silica gel, a facility to fill the eluent on the top allowing to percolate through the silica gel, and the possibility to collect the eluent leaving the column at the bottom. As well, the selection of the polarity of the eluent is the dominant parameter determining the quality of the separation. For complicated separation problems, e.g. isolation of an analyte from various chemically very similar substances (as purification step), one solvent or solvent mixture is used with a very fine collection of fractions. Coarse fractionations of a wider range of polarity, e.g. for separating roughly the broad spectrum of compounds in raw extracts, are carried out by changing the eluent composition and to collect the fractions according to the different eluents.

In organic-geochemical analyses of fossil materials, a coarse separation of crude extracts or oils into three fractions is used. The individual fractions represent mainly different compound classes, the aliphatic and aromatic hydrocarbons as well as the functionalized compounds, also called NSO-compounds. For environmental analyses a broader fractionation can be used as illustrated in Fig. 3.21. But as well, the separation of individual compounds or substance classes is aimed at.

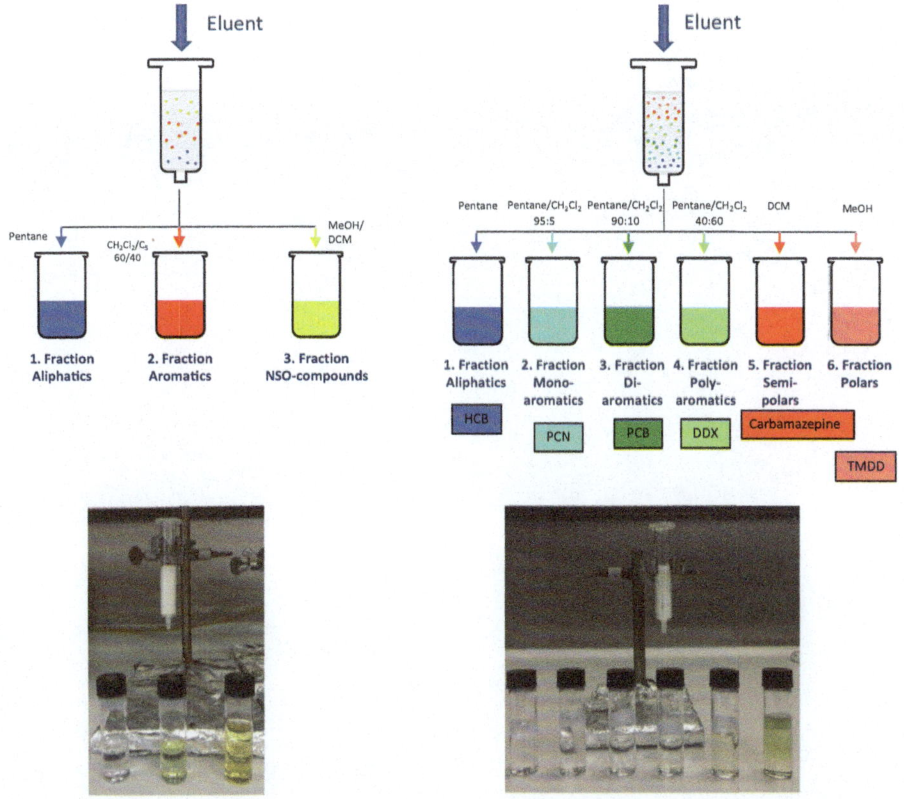

Fig. 3.21 Typical fractionation scheme for fossil material (e.g. coals) and environmental extracts (e.g. from sediment or soil samples, examples of corresponding pollutants are given)

General Note
Successful analyses of crude extracts and complex mixtures of analytes need a fractionation step. This procedure is mainly done by low-performance liquid column chromatography resulting in several subfractions containing different substance classes with increasing polarity.

References

Cammann K (2010) Instrumentelle Analytische Chemie. Spektrum Akademischer, Heidelberg. isbn:978-3-8274-2739-7

Schwedt G (2007) Taschenatlas der Analytik. Wiley-VCH, Weinheim. isbn:978-3-527-31729-5

Chapter 4
Instrumental Analysis

4.1 High Performance Chromatography: GC, HPLC

Outlook
Gas chromatography GC and high-performance liquid chromatography HPLC are fundamental analytical techniques. Here, their technical realization and the basic parameters for optimizing the chromatographic separation are presented. However, also restrictions and pitfalls are mentioned.

4.1.1 Gas Chromatography GC

Gas chromatography is one of the by far most important instrumental techniques used in Organic Geochemistry. Its importance increased especially since a direct online coupling with mass spectrometry (GC/MS) has been enabled. As already introduced in Sect. 3.3.1, the basic concept of GC comprises gas as the mobile phase and a fluid stationary phase. Modern GC systems are able to separate analytes with an extraordinary separation efficiency and on low concentration levels on the ng-or even pg-level.

A gas chromatograph, the technical realization of gas chromatography, exhibits a relatively simple set-up (see Fig. 4.1). It consists mainly of three parts, the injector, the detector and the main chromatographic component, the column. The column is placed within an oven. In principal, the analytes are introduced via the injector and become separated by flowing together with the carrier gas through the column. At the end of the column, a detector is recording compounds in the effluent qualitatively and quantitatively.

The core of a GC system is the column. High-performance gas chromatography is using capillary columns with very narrow diameters and high length (see Fig. 4.2).

J. Schwarzbauer, B. Jovančićević, *Introduction to Analytical Methods in Organic Geochemistry*, Fundamentals in Organic Geochemistry,
https://doi.org/10.1007/978-3-030-38592-7_4

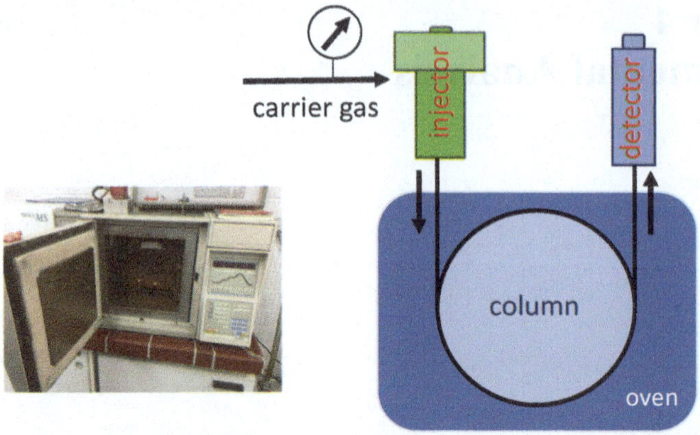

Fig. 4.1 Principal composition of a gas chromatograph

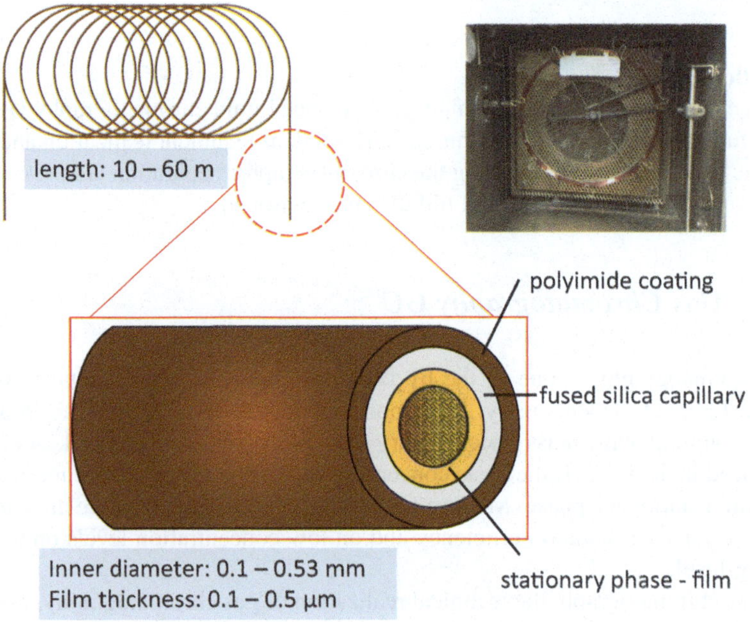

Fig. 4.2 Dimension and structure of a typical capillary GC column

Formerly, also short columns with broader diameters have been used, but nowadays they are more or less completely replaced by the capillary technique. The typical dimensions of a capillary column cover a length between 10 and 60 m and a diameter between 0.1 and 1.0 µm. Capillary columns consist of glass coated by a polyimide film. This polymer film makes the column flexible and prevents a breaking of the glass. This technical feature allows to wind the long column on carriers with diameters of 10–20 cm enabling an easy handling of the long columns and a positioning of them in an oven.

In gas chromatography normally a liquid phase is used as stationary phase. In order to obtain an immobilized fluid, not a true liquid but a pseudo-liquid is utilized. For this purpose, polymer molecules are fixed (mostly chemically) at the inner wall of the column. Due to its length and the resulting mobility similar to liquid molecules, these polymer films act physically similar to fluids. The retention interaction of the by far highest proportion of GC columns used in Organic Geochemistry is based on the partition between liquid and gas phase. Hence, the chemical properties of the film determine the retention behavior. A major part of GC columns is equipped with polysiloxane based films. Chemical derivatization of the polysiloxanes with aliphatic moieties (e.g. methyl groups) produce nonpolar films, insertion of more polar moieties such as nitrile groups or phenyl rings enhance the polarity, whereas films composed of ethylene glycol moieties represent very polar films (see Fig. 4.3).

Dimension and type of film are highly influencing the gas chromatographic separation. The thicker the film, the higher the retention and, as a consequence, the retention time but also the separation capacity increases (see Fig. 4.4). Changing the type of film influences the separation capacity but also the retention order of the individual substances. This becomes visible especially for compounds of different chemical properties as illustrated in Fig. 4.4. On a similar way, the dimension of the column systematically alters the separation efficiency but also the analysis time. A summary of the quantitative impact of column length and diameter is given in Table 4.1.

Noteworthy, the huge success of gas chromatographic analyses is related also to a second interaction superimposing drastically the already introduced partition processes. In former times, gas chromatographic separations have been performed at a constant temperature (isothermal), but in modern times a ramped temperature program is applied during the separation process. As a result, a type of distillation is additionally overlaid. Hence, the separation depends on both boiling point and polarity of the analytes. The influence of temperature on the separation efficiency is exemplified in Fig. 4.5. Typical temperature programs start in the region of solvent boiling points (50–80 °C), have a short isothermal time around 2–5 min and ramp following the temperature to endpoints around 300–350 °C with a heating rate between 2 °C and 15 °C/min. At the end, the final temperature is hold for several minutes. Taking such common values in mind, analysis times between 30 and 100 min are typical.

Beside the stationary also the mobile phase is an important element of chromatography. In contrast to liquid chromatography, the role of the mobile gas phase in GC is less important. One key aspect is the demand that the carrier gas must not react with the analytes or the solvent. Hence, only chemically inert gases can be used and, therefore, the spectrum of gases is very limited comprising solely hydrogen nitrogen and helium. For these gases one parameter has an influence on the separation efficiency, the carrier gas velocity. Theoretical description of this influence is given by the Van-Deemter equation, where the term H (separation step, an inverse

common phases for gas chromatography		
OV1, DB-1, SE30, CP-Sil 5, ZB1	Permethylated polysiloxane	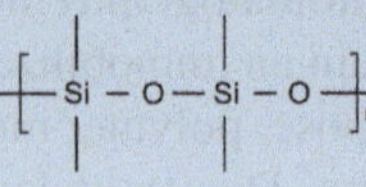
BPX5, SE52, DB5, SPB-5 , ZB5	Permethylated polysiloxane with 5% phenyl substitution	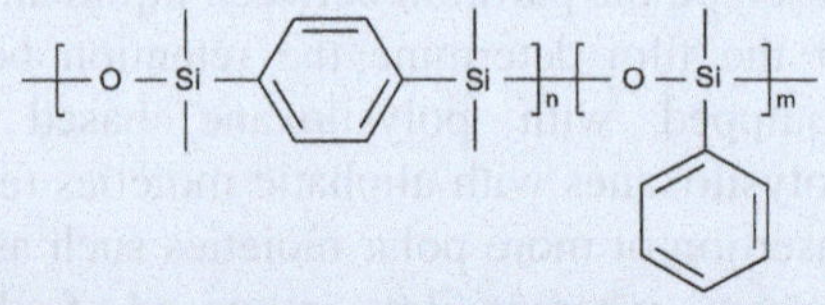
SE54	Permethylated polysiloxane with 5% phenyl and 1% vinyl substitution	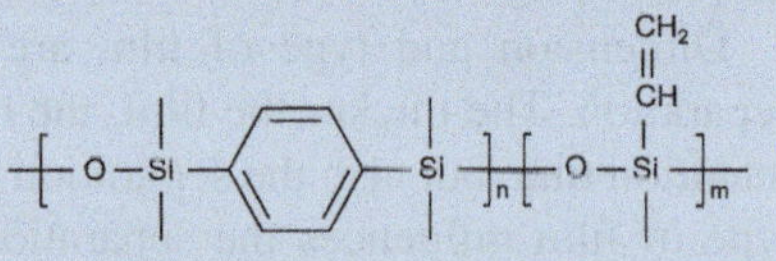
OV1701, DB5, SPB-7, CP-Sil19	Permethylated polysiloxane with 7% phenyl and 7% cyano substitution	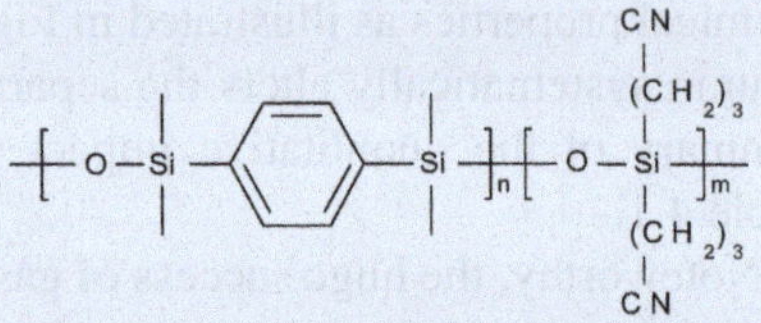
DBWax, CW20M, BP-20	Polyethylene glycol (MW 20000)	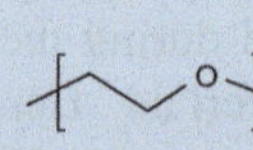
FFAP	Polyethylene glycol (MW 20000) partially esterified with 2 - nitro terephthalic acid	

Fig. 4.3 Chemical composition of some typical films used in gas chromatography

parameter to separation efficiency!) is described by three additive terms. Two terms are influenced by the linear gas velocity, one with a linear correlation, one with an opposite working inverse correlation. As a consequence, the equation exhibits a minimum representing a maximum of separation efficiency as illustrated by the Van-Deemter graphs (see Fig. 4.6). This implies different ideal gas velocities for the

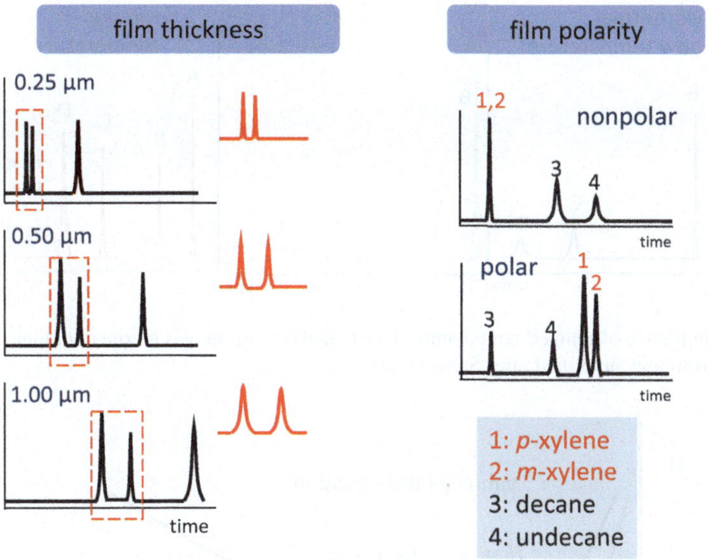

Fig. 4.4 How film type and dimension influence the gas chromatographic separation

Table 4.1 Influence of column dimensions on chromatographic parameters

	Length (L)	ID (r)	Phase thickness (d_f)
Resolution (R)	$R \sim \sqrt{L}$	$R \sim \frac{1}{r}$	$R \sim \frac{1}{\sqrt{d_f}}$
	To double the resolution, quadruple the length	Resolution decreases with increased diameter	Resolution decreases with increased thickness
Sample capacity (SC)	$SC \sim \sqrt{L}$	$SC \sim r^2$	$SC \sim d_f$
	Longer columns have a slightly better capacity	Capacity is exponential as diameter increases	Sample capacity increases with thicker films
Retention time (t_r)	$t_r \sim L$	$t_r \sim r$	$t_r \sim d_f$
	Longer columns require longer analysis times	Smaller diameter columns allow faster analysis times	Thicker films require longer analysis time

carrier gases: helium should be used with lower velocity as compared to hydrogen but higher ones as compared to nitrogen. This has certainly also a direct impact on the analysis time.

General Note
Modern GC analysis uses capillary columns and measurements with temperature programs to achieve high-performance separation.

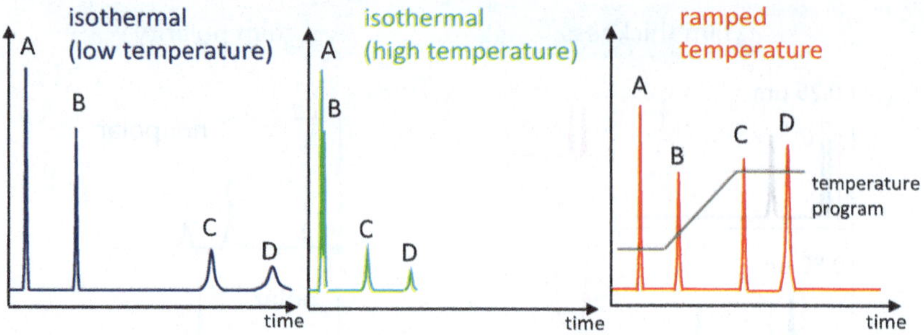

Fig. 4.5 Influence of ramped temperature during analysis on the gas chromatographic separation (adopted from and simplified after Schwedt 2007)

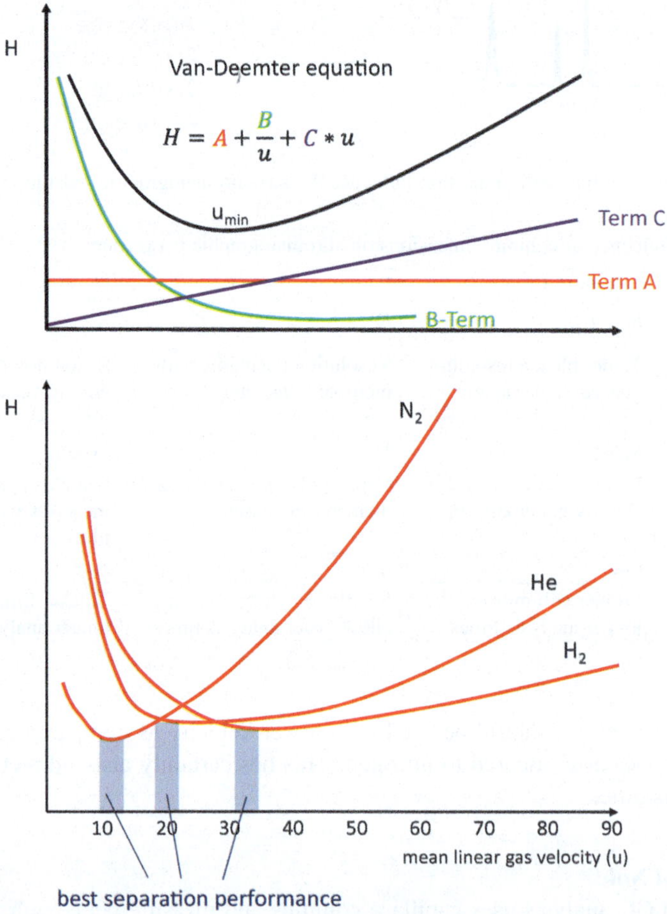

Fig. 4.6 The role of carrier gas velocity on separation efficiency in gas chromatography as described by the Van-Deemter relation

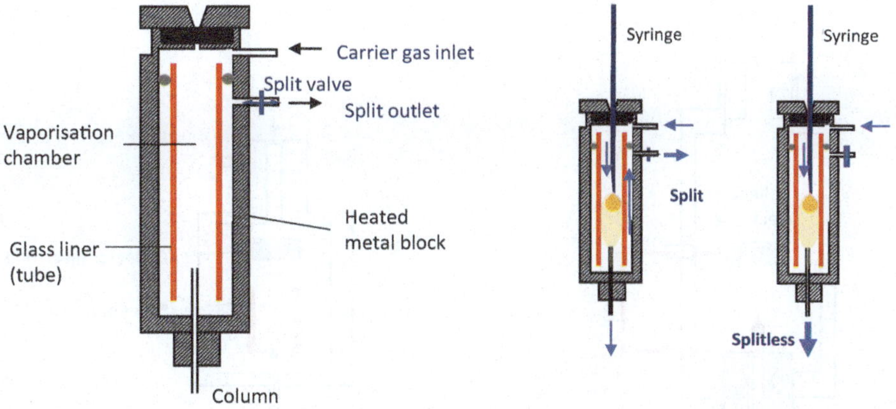

Fig. 4.7 Scheme of a split/splitless injector and the two principal injection modes

For GC analysis not pure extracts or substances are used, but solutions at low concentration levels. Further on, the analytes need to be transferred to the gas phase prior to separation. Hence, an important part of a gas chromatography is the system transferring the analytes form a dissolved state in a solution to gaseous state and, subsequently, towards the capillary column. This function is performed by the injector. Nowadays the split/splitless injector type is mainly used. A principal scheme is given in Fig. 4.7. Core of this system is a small glass tube that is heated to 250–300 °C. The carrier gas is flowing through the tube from top to bottom. One end of the capillary column is placed at the bottom. The carrier gas overpressure determines its flow rate through the capillary column. Above the top of the glass tube a septum is installed that allows a penetration by a syringe. For the injection a small volume (normally 0.2–1 μL) in a syringe, which tip is placed through the septum in the middle of the tube, is sprayed within the hot tube space. The solvent and the analytes get rapidly vaporized and are transported by the gas flow onto the column thereafter. A special feature of this type of injector is the possibility to split the gas flow. For this purpose, a valve allows a major part of the carrier gas to leave the injector without passing the column. This induces a split of the gas flow between a proportion passing the column and a much higher fraction that gets lost. Typically, the split ratio of column passing fraction and the bypassed fraction is around 1:50. However, the effect is on the one hand to avoid an overloading of the column especially by the solvent but more important a discrimination takes place. The very volatile solvents are preferentially following the bypass, but the less volatile analytes are more concentrated on the capillary column. Hence, a slight separation of analytes and solvents is achieved by this technique. If concentrations of analytes are very low, the split mode hinders a transfer of its total amount reducing the sensitivity of the gas chromatographic measurement. Here, an alternative spitless mode can be used by closing the split valve during injection forcing the total injected sample volume to enter the capillary column. The difference between split and splitless mode injection is illustrated in Fig. 4.7.

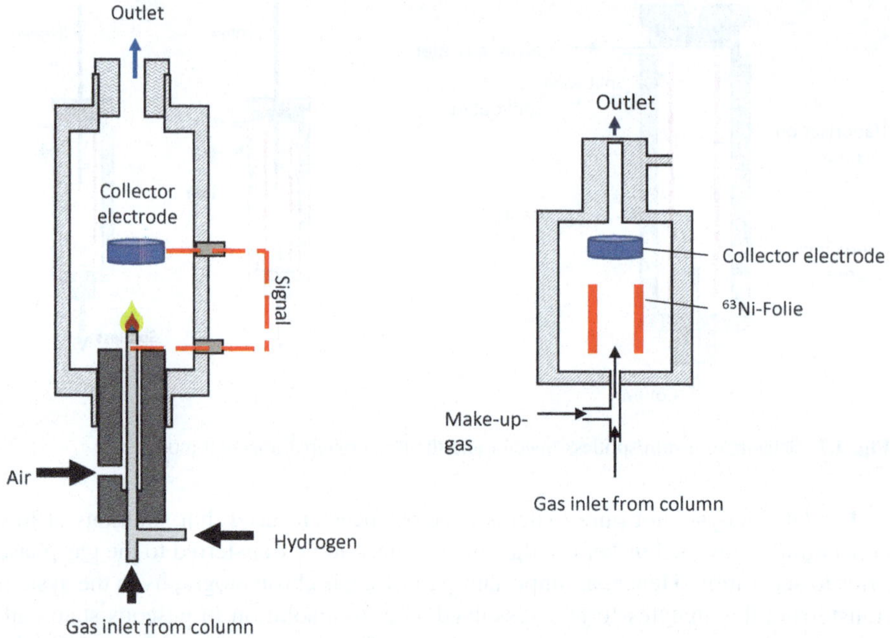

Fig. 4.8 Schemes of a flame ionization detector FID (left) and a more selective electron capture detector ECD (right)

Beside the standard injector various technical variations are established, e.g. allowing the injection of higher volumes or with a very fast ramped temperature within the injector (e.g. PTV injectors). Also injectors allowing a direct positioning of the analyte solutions in the capillary column are used (e.g. cold-on-column injectors).

General Note
As a highly relevant aspect, the discussion of the injection modes clearly points to a major restriction of gas chromatography. It is limited to analytes that can be vaporized under the conditions described for the injectors. Hence, only high to medium volatile substances can be analyzed.

As third main part of a gas chromatograph, the detector has various important properties influencing the quality of GC measurements. Also here, one type is dominant, the flame ionization detector, FID. Its technical basics are illustrated in Fig. 4.8. A detector converts a substance leaving the capillary column into an electronic signal. Thereby, information about the appearance of a substance and its quantity can be recorded. For this purpose, the FID uses the ionization processes that occur if organic molecules stay for a short time stay in a flame. In the FID the eluting substances are passing an oxyhydrogen flame and become ionized. Due to the gas flow the ions are transported out of the flame and pass as second step a collector electrode. The interaction of the moving ions (electric charge) with the voltage field of the electrode induces an electronic signal. This signal is highly sensitive allowing the detection of very low amounts of organic substances (down to pg levels) but excludes inorganic species like the carrier gas or water. Further on, the detection is not specific or selective. More or less all organic compounds can be detected, but noteworthy with different sensitivity. The latter point has a high relevance for quantitation (see Sect. 5.2). As second important aspect for quantitative measurements, the linear correlation of amount of compound and electronic signal, the so-called linear range, should cover a wide extent. The good sensitivity, its wide linear range and the broad spectrum of detectable substances are the main reasons for the dominant application of the FID.

For some more specific applications, more selective detectors are used. The principal approach is always the same, organic substances are converted to an electronic signal by ionization. However, the type of ionization changes and, simultaneously, also the selectivity due to different ionization potentials for individual substance groups at softer ionization processes as compared to flame ionization. As an example, ionization by β^--radiation is performed by the electron capture detector (ECD) via the ^{63}Ni-isotope (see Fig. 4.8). This soft ionization allows the selective detection of halogenated compounds as well as sulphur-containing compounds but exhibits a poor sensitive for nonhalogenated organic substances. Therefore, this detector is applied especially in environmental analysis measuring e.g. dioxine or PCB pollution. Further specific detectors are e.g. the NPD (nitrogen phosphor detector) used in particular for pesticide analyses or the photo ionization detector (PID) applied for PAH detection.

A systematic comparison of some commercially available detectors regarding the sensitivity, linearity and selectivity is given in Fig. 4.9.

As already mentioned, gas chromatography is restricted to volatile compounds. Since volatility of organic substances depends mainly on the weight and the polarity of a compound, high molecular weight and polar substances are excluded. However, the discrimination between detectable and non-detectable it is not sharp, in particular for moderate polar substances such as carboxylic acids or mono alcohols. Many of these compounds are slightly volatile but have a poor separation performance resulting in bad peak forms such as tailing or fronting. In order to get these compounds better detectable, derivatization is used. Derivatization covers chemical reactions leading to less polar compounds by conversion of polar functional groups into less polar ones. This approach is dominantly applied for carboxylic acids and alcohols that can be transferred to less polar esters or silyl ethers, respectively. Derivatisation agents are added to the analytes prior to analysis and comprise substances such as diazomethane, methanol/borone trifluoride solution, TMSH for ester formation or MSTFA, BSTFA for silyl ether generation. Examples are illustrated in Fig. 4.10.

Name	Abbreviation	Selectivity	Detection limit	Linearity range
Flame ionisation detector	FID	universal	low pg C	> 10^6
Electrons capture detector	ECD	selective (halogens, S)	strongly compound depending (down to fg)	10^4
Thermal conductivity detector	TCD	universal	Low ng	10^5
Flame photometry detector	FPD	selective (S, P)	low pg S high fg P	10^3 10^4
Nitrogen/phosphor detector	NPD	selective (N, P)	high fg N,S	10^5

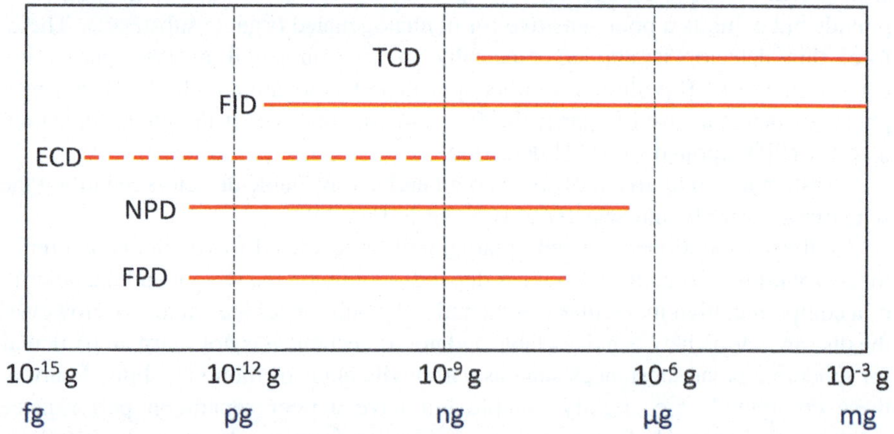

Fig. 4.9 Parameters of different GC detectors and an illustration of the detector sensitivities and linearity ranges (partly modified after Cammann 2010)

General Note
Derivatization can be used for extending the spectrum of GC detectable substances by lowering their polarity and enhancing volatility as well as peak performance.

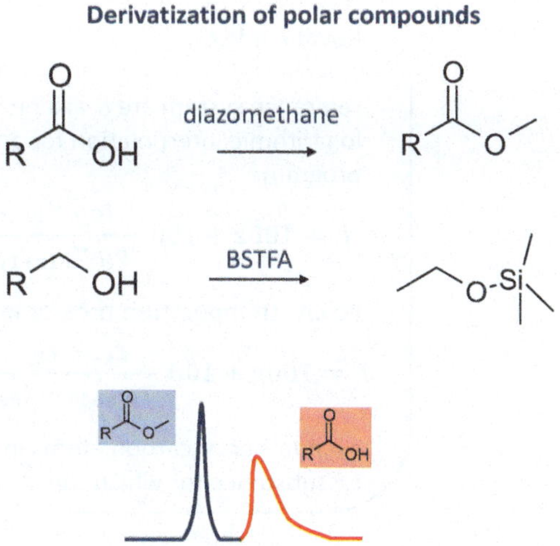

Fig. 4.10 Derivatization of carboxylic acids forming methyl esters or alcohols forming silyl ethers for a better gas chromatographic measurement as exemplified by the peak form of an acid and the corresponding methylester

Finally, gas chromatography is a separation method that can provide first evidence on compound identification. The retention behavior is a characteristic property and can be used for characterization of individual substances. However, the absolute retention time is not suitable since it varies depending on the machine, the gas velocity, temperature program, column length and many other parameters. But using a same type of column film (e.g. methyl siloxanes like ZB1, HP1 or OV1, see Fig. 4.3) the relative retention and the corresponding retention order is always the same regardless of all other parameters. This phenomenon has been used to define so-called retention indices that can be used for compound characterization instead of retention times.

Two index systems are well-known, the Kovats index and the Lee index. They differ in one basic aspect, they have a different frame of defined reference index values. The Kovats index is related to the retention of *n*-alkanes, where each *n*-alkane gets the index composed by the number of carbon atoms multiplied by 100. The retention index of another substance can be calculated by interpolation of retention times of both substance and reference alkanes related to retention index of reference alkanes. Both normal or logarithmic interpolation is used for isothermal or temperature programed measurements, respectively. This is illustrated in Fig. 4.11. Main advantage of this method is the possibility not only to interpolate but also to extrapolate to some extent from given *n*-alkane values due to the very regular retention behavior of the reference compounds, the *n*-alkanes. To be accurate, extrapolation over the whole range of a chromatogram is only valid for an

Kovats Index

reference system are *n*-alkanes;
logarithmic interpolation for an isothermal temperature program:

$$I_i = 100z + 100\,\frac{\log t'_{Ri} - \log t'_{Rz}}{\log t'_{R(z+1)} - \log t'_{Rz}} \qquad t'_{Rz} < t'_{Ri} < t'_{R(z+1)}$$

Kovats (temperature programmed):

$$I_i = 100z + 100\,\frac{t_{Ri} - t_{Rz}}{t_{R(z+1)} - t_{Rz}}$$

z = number of carbon atoms in reference *n*-alkanes
i = substance for which the index shall be calculated

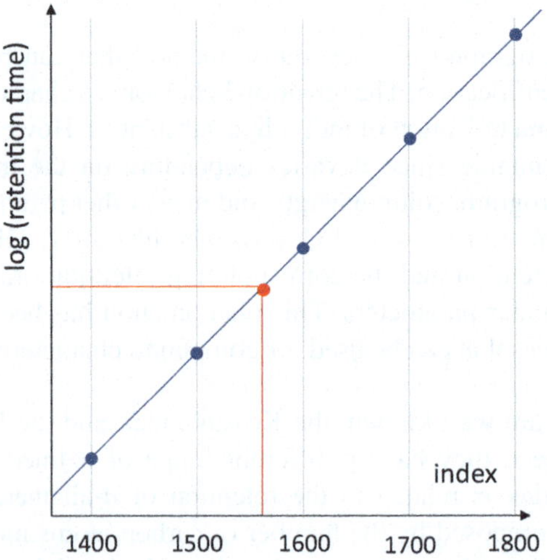

Fig. 4.11 The concept of the Kovats index

isothermal measurement with logarithmic interpolation. However, in a more narrow retention time range, extrapolation is also applicable for temperature-programed measurements with linear interpolation.

As a second index system, the Lee index uses aromatic compounds as reference frame, whereby the number of rings defines the Lee index multiplied by 100. Benzene and naphthalene as one and two ring aromatic hydrocarbons get the Lee indices 100 and 200, respectively. For higher ring numbers one representative each has been selected such as phenanthrene (three rings), chrysene (four rings), picene (five rings) etc. For Lee indices of other compounds, a linear interpolation is used according to the temperature programed Kovats index (see Fig. 4.12). Since the

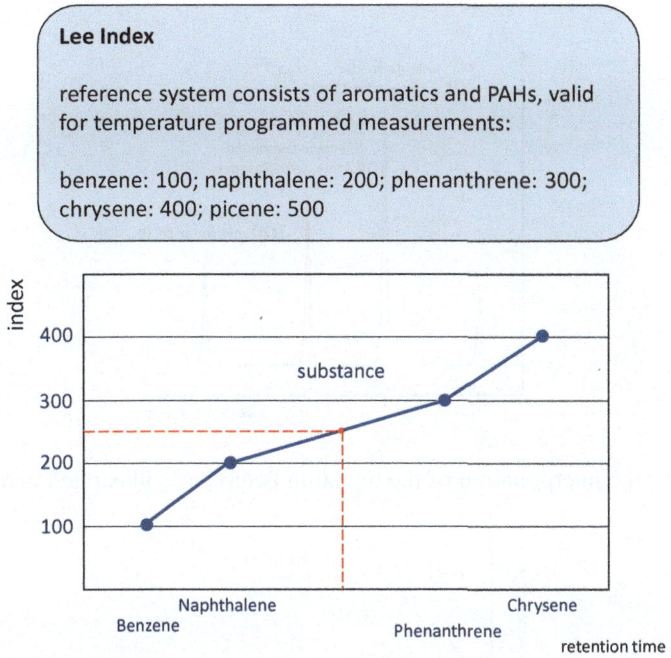

Fig. 4.12 The concept of the Lee index

relative retention of these reference substances is less regular, only a strict interpolation but no extrapolation is allowed.

General Note
Providing the same column film, the GC retention index of a given substance is always the same and, consequently, can be used as first evidence for compound identification.

Case example: ***How to calculate a Kovats index***
This example exemplifies how to calculate a Kovats index for given gas chromatographic data of an isothermal measurement. The measured data are as follows:

(continued)

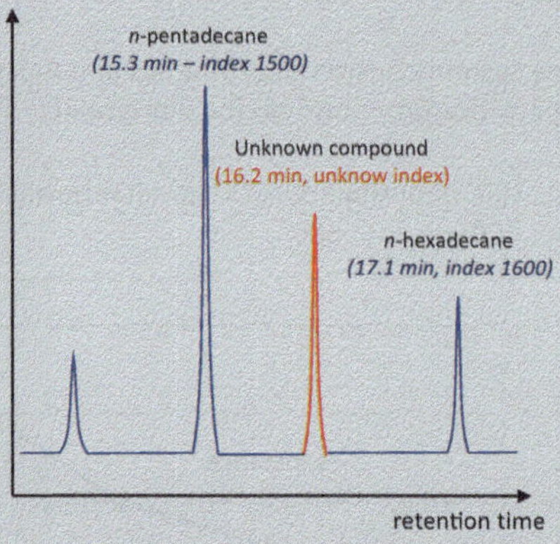

A graphic interpretation of the retention behavior is illustrated here:

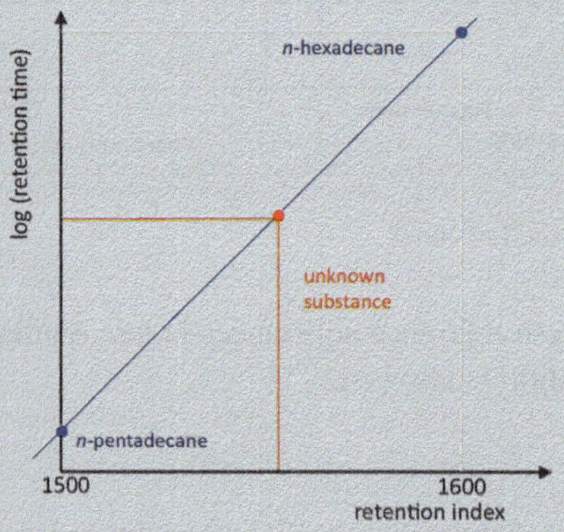

However, applying the logarithmic interpolation given by equation given in Fig. 4.11 results in:

$$I_{unknown} = 1500 + 100\,\frac{\log 16.2\ - \log 15.3}{\log 17.1\ - \log 15.3}$$

or

$$I_{unknown} = 1500 + 100 \bullet 0.5139 \approx 1551.4$$

The calculation reveals a Kovats index of approx. *1551* for the unknown compound.

4.1.2 *High Performance Liquid Chromatography HPLC*

As already mentioned, gas chromatography has a major restriction, it is only applicable on volatile substances. Compounds with elevated polarity (e.g. glucose) exhibit normally high boiling points or even just decomposition points. Hence, they cannot be vaporized under normal conditions and, consequently, cannot be analyzed by GC. The same accounts for high molecular weight compounds such as biomacromolecules (e.g. cellulose) or synthetic polymers (e.g. polystyrene).

An alternative separation technique for these compound groups is the liquid chromatography, normally performed as high-performance liquid chromatograph HPLC. The importance of HPLC increased enormously end of the 90s, because the first well working systems linking HPLC and mass spectrometry (LC/MS) became commercially available.

HPLC systems are built up similar to GC systems, exhibiting an injection system, the chromatographic column and various detector systems (see Fig. 4.13). However, the mobile phases are liquids and, therefore, the restricting criterion for the useful application of HPLC is the solubility of the analytes in the mobile phase. The usage of liquids as eluents implies two requirements. Firstly, for a successful flow of the eluents through the columns packed with solid particles an enhanced pressure (up to 350 bar) is needed. Secondly, some differences in the dimensions of the columns exist as compared to GC. Lengths of HPLC columns are shorter (commonly 15–25 cm), but the column diameter is wider (usually 4.6 mm). Over the last decades

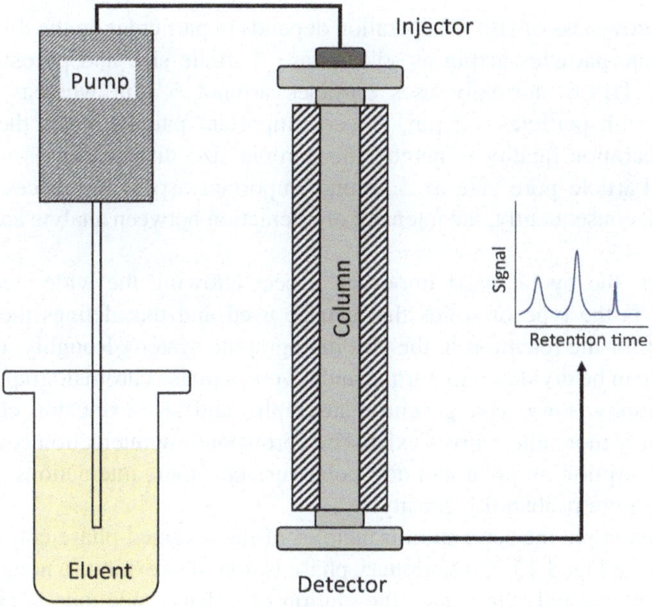

Fig. 4.13 Principal construction of a HPLC system (adapted and simplified after Schwedt 2007)

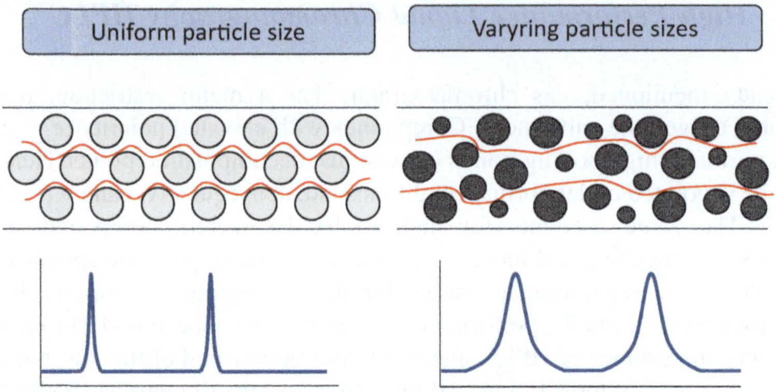

Fig. 4.14 Optimization of HPLC separation by particle size distribution of the stationary phase

a trend towards smaller columns became obvious resulting in ultra-high-performance liquid chromatography UHPLC using columns with diameters down to 2 mm but also higher pressure (up to 1200 bar).

General Note
HPLC represents a chromatographic method highly complementary to GC. It allows to detect those substance classes that are not accessible for analysis by GC—the polar compounds and the higher molecular weight compounds.

The effectiveness of HPLC separation depends in particular on the dimension and quality of the particles acting as solid phase. Particle size and porosity are main parameters. HPLC normally uses particles around 5 μm, whereas UHPLC is performed with particles <2 μm. A very important parameter for the chromatographic separation quality is hereby the particle size distribution as illustrated in Fig. 4.14. Particle pore size as a second important aspect influences the overall surface and, consequently, the intensity of interaction between analyte and stationary phase.

However, the by far most important aspect allowing the wide area of HPLC application is the type of solids that can be used and that defines the interaction responsible for the retention in the chromatographic system. Roughly, the different techniques can be divided into normal and reversed phase chromatography, partition chromatography, ion-exchange chromatography and size-exclusion chromatography, but many more alternatives exist. The corresponding interaction covers adsorption and desorption on polar and non-polar surfaces, ionic interactions, partitioning and particle/pore probability density.

As one example the separation principles of the reversed-phase chromatography is illustrated in Fig. 4.15. If a stationary phase is used that exhibits a nonpolar surface (e.g. peralkylated polysiloxanes), the elution of polar compounds is facilitated by more polar eluents due to the lower interacting forces between analyte and stationary

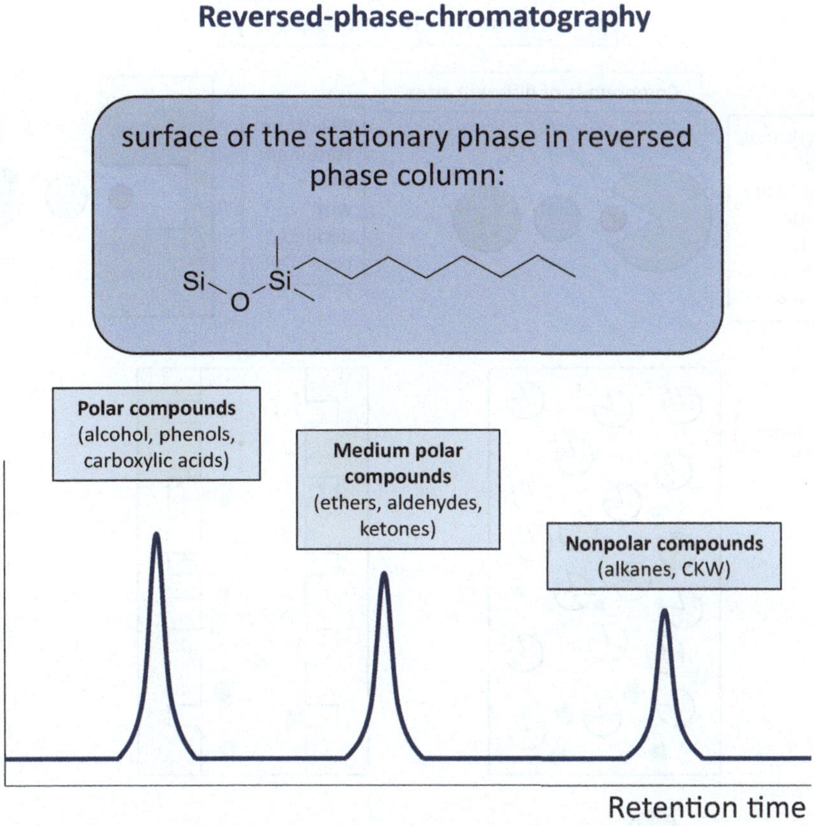

Fig. 4.15 Principle elution order in reversed-phase chromatography

phase but favored interactions between analyte and mobile phase. As a consequence, the elution follows the order polar > medium polar > nonpolar. Since this elution order is inverse to the common chromatographic systems established since many decades in low pressure column chromatography (using polar stationary phases e.g. silica gel or Al_2O_3 and more nonpolar eluents), this chromatography is named 'reversed-phase' chromatography. Common phases in reversed-phase chromatography are C_8 or C_{18} phases, consisting on polysiloxanes with octyl or octadecyl substituents.

As a second example, the separation process used in size-exclusion chromatography is figured out in Fig. 4.16. Here, the stationary phase exhibits particles with pores that cover a wider range of pore sizes. Only the average duration of stay within the stationary phase is responsible for a different retention of particles with different size. All other interactions are suppressed by selecting a suitable eluent. The retention is related to the possibility for analytes to enter and leave the pores. Small particles can enter a much wider range of pore sizes (and consequently a higher number of pores) and need, therefore, more time to pass the column. Big

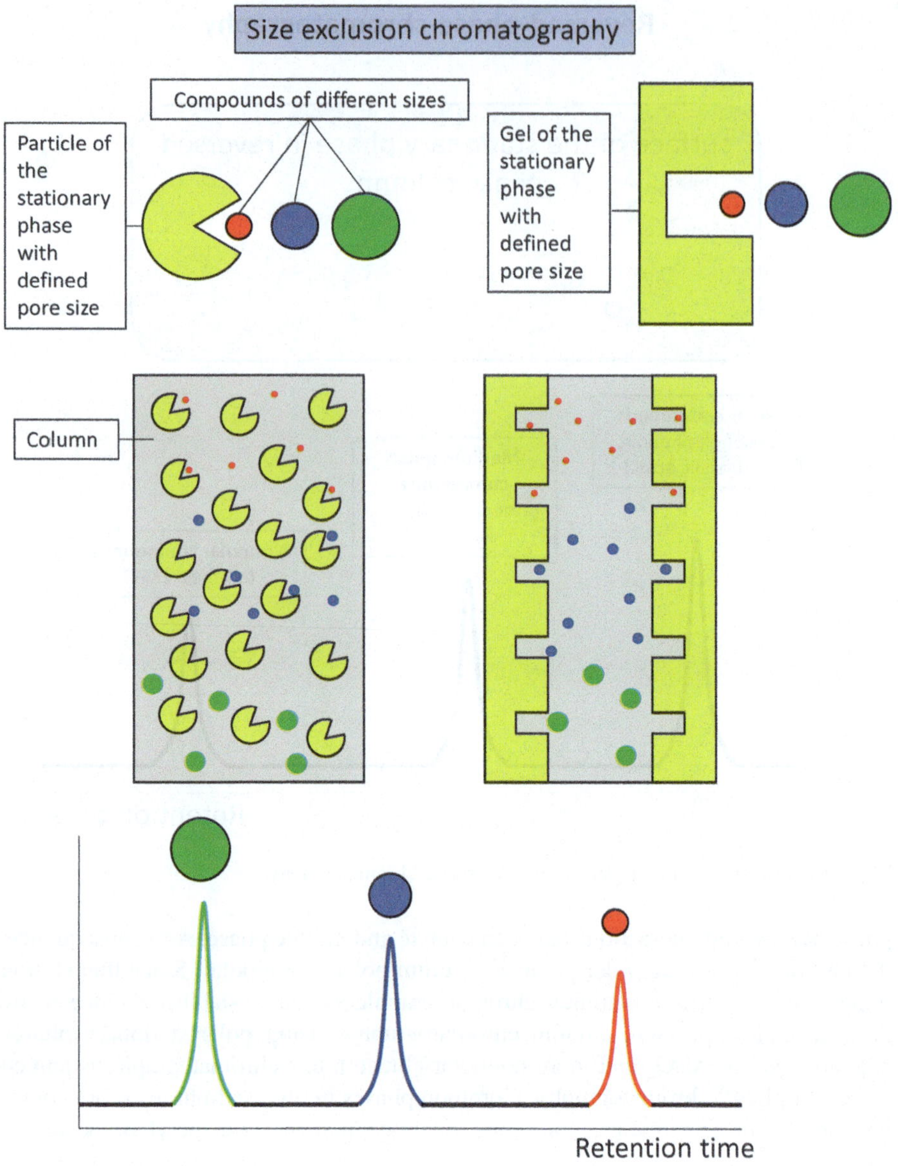

Fig. 4.16 Principles of size-exclusion chromatography

particles are only enabled to enter the bigger or even biggest pores, hence their residence time within the column is shorter. This implies an elution order of high molecular weight analytes > medium molecular weight analytes > low molecular weight analytes. The pore sizes and their distribution determine the range of molecular weights that can be separated on the individual columns.

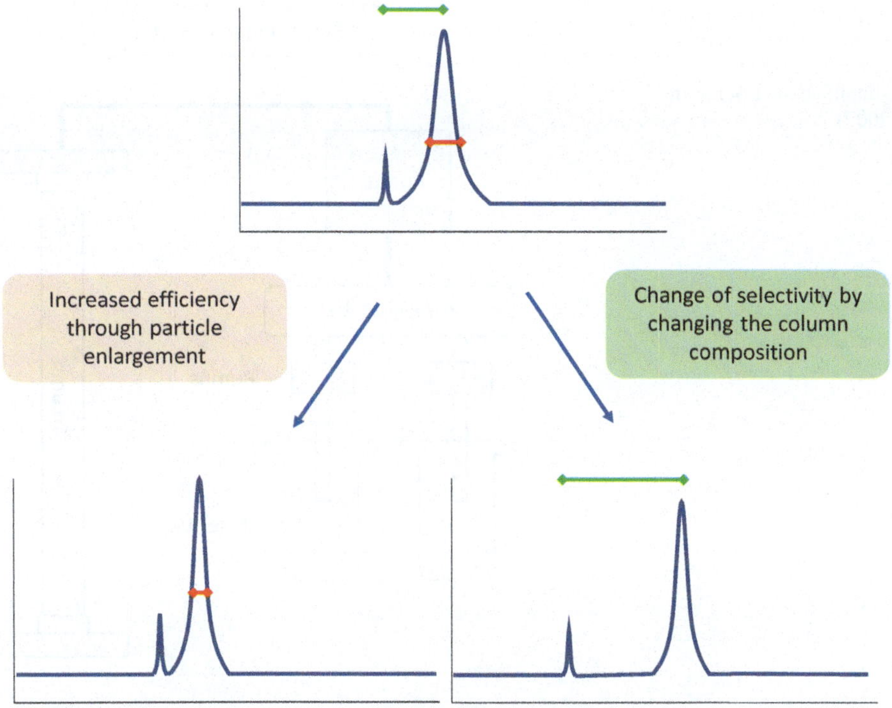

Fig. 4.17 Influence of particle dimension parameters and type/selectivity of stationary phase on the chromatographic separation

The selection of appropriate stationary phases has a high influence on the chromatographic separation, partly much higher as compared to particle size, particle pore size or similar parameters. This is illustrated in Fig. 4.17. Comparable to gas chromatography, mainly the peak width and the retention efficiency are influenced.

As a final aspect, there is a last aspect that can be used to optimize HPLC based separation. Initially, HPLC has been performed with solely one eluent, a so-called isocratic elution. In particular for the reversed-phase chromatography, the application of a changing eluent composition has been applied very successfully. This approach is called gradient elution. Normally only two or three different eluents with different properties, e.g. in polarity, are used representing binary or tertiary gradient systems (see Fig. 4.18). The change of composition is a systematical gradient but does not need to be linear as illustrated in Fig. 4.18. A variety of non-linear gradients can be used.

The effect of a gradient separation as compared to an isocratic one can be seen in Fig. 4.19. The peak width is improved by contemporary shortening of elution time. These effects are similar to improvements achieved by temperature programed GC analyses as compared to isothermal analyses (see Fig. 4.5).

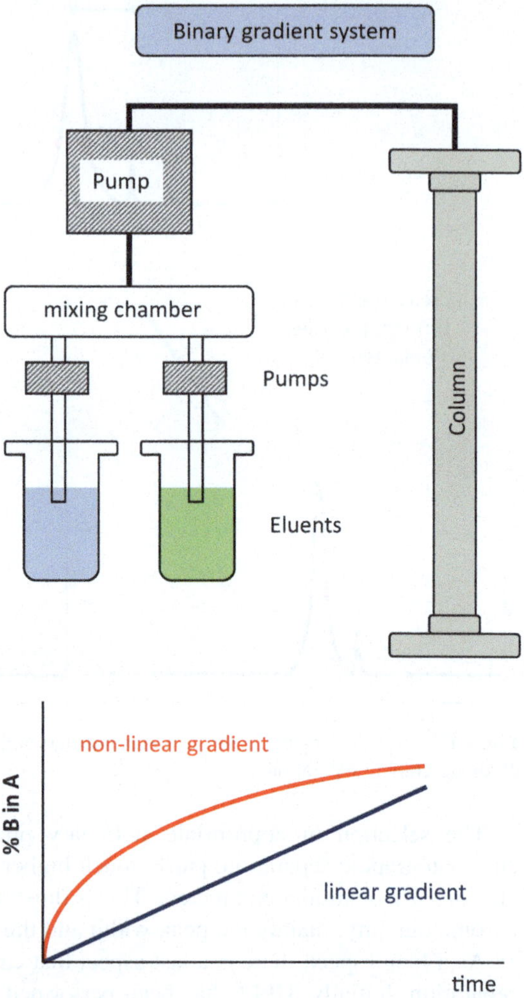

Fig. 4.18 Binary gradient system as used in HPLC (adopted from and simplified after Schwedt 2007)

General Note
For optimization of HPLC separation several different parameters can be used: particle size and porosity, variation of interaction and forces between analytes and stationary phase, gradient elution and some more. All parameters influence chromatographic resolution and efficiency.

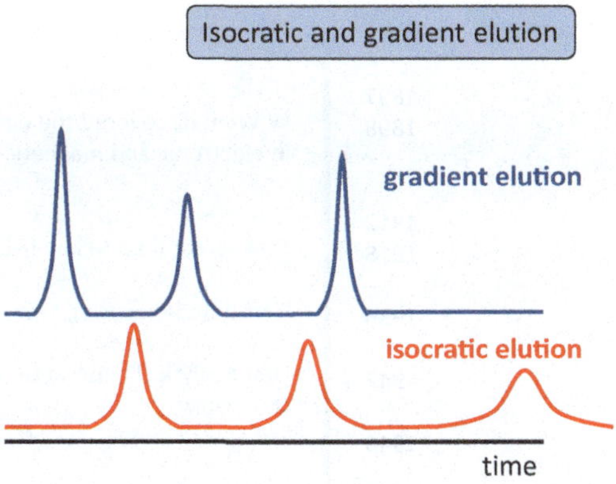

Fig. 4.19 How a gradient elution improves the HPLC based separation (adopted from and modified after Schwedt 2007)

4.2 Mass Spectrometry MS

Outlook
Mass spectrometry is the most important method for identification of organic compounds in Organic Geochemistry. In combination with gas chromatography (as GC/MS) it is also the basic method for quantification. Fundamental technical aspects and applications are presented and discussed here.

Mass spectrometry (MS) is the by far most important instrumental analysis technique in Organic Geochemistry. Its importance is based on two major characteristics: mass spectrometry is fast and very sensitive. The fast measuring allowed at a very early stage (mid of the twentieth century) to link mass spectrometry with the very efficient separation technique of gas chromatography forming so-called GC/MS systems. In addition, the high sensitivity of detection fits ideally the sensitivity of capillary GC. As a consequence, these sensitive GC/MS systems opened a wide field for analyzing organic compounds in complex mixtures and at low concentration levels.

Excursus: ***Brief historical development of mass spectrometry***
As all other analytical techniques mass spectrometry has been developed over decades. To get an impression of this process, a brief historical summary is given below that points to some important steps in mass spectrometry from the discovery of the fundamental physical processes towards modern technical set-ups:

(continued)

1897	JJ Thomson, determined the m/z ratio of an electron
1898	W Wien discovered the deflection of charged particles in electronic and magnetic fields
1912	Mass spectrometric separation of neon isotopes
1918	First sector field MS by JF Dempster
1936	First double focusing sector field MS
1942	First analytical application in petroleum (organic) chemistry
1946	Time-of-flight MS, ToF MS, by WE Stephens
1953	Quadrupol MS by W Paul
1954	JH Beynon established MS based structure elucidation of organic molecules
1960	Systematic studies on electron impact induced fragmentation in MS sources
1966	Chemical ionization, CI
1984	Electrospray ionization ESI

Please note, this time line does not claim to be exhaustive.

4.2.1 Principals of Mass Spectrometry

Basic physical principal of mass spectrometry is the interaction of charged particles with electric or magnetic fields that induce either a linear acceleration or a forcing on a circuit as the result of radial acceleration. These accelerations are depending on the mass to charge ratio (m/z) of the particles. This interrelation is used to detect different masses of elements but also of molecules. The later application is used in organic mass spectrometry.

Keeping this basic relation in mind, the principle composition of a mass spectrometer becomes obvious (see Fig. 4.20). First a unit generating ions is needed, the so-called ion source, followed by a second device that performs the separation of the charged particles. Finally, a detector is needed counting the separated ions.

Detecting ions is very simple, performed by multiplier detectors. However, the ion source as well as the separator unit are the more important parts of mass spectrometer. The technical realization of ionization of organic molecules depends on the type of transfer or injection of the analytes. A direct injection is not common in organic-geochemical analyses but, as already mentioned, the linkage with gas chromatography or liquid chromatography is used. There is one important

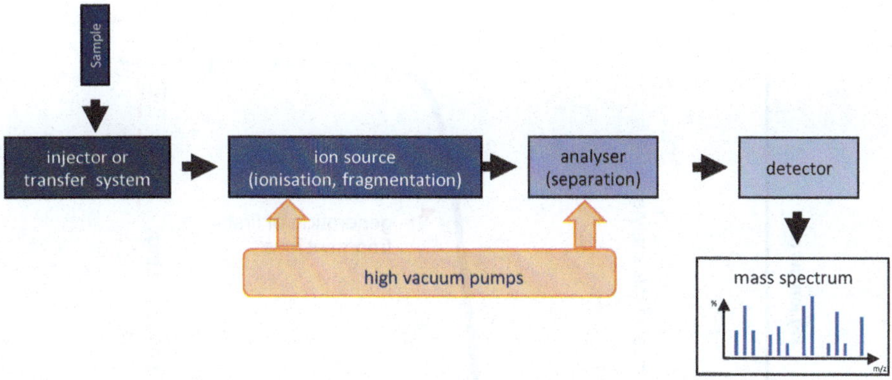

Fig. 4.20 Principle structure of a mass spectrometer

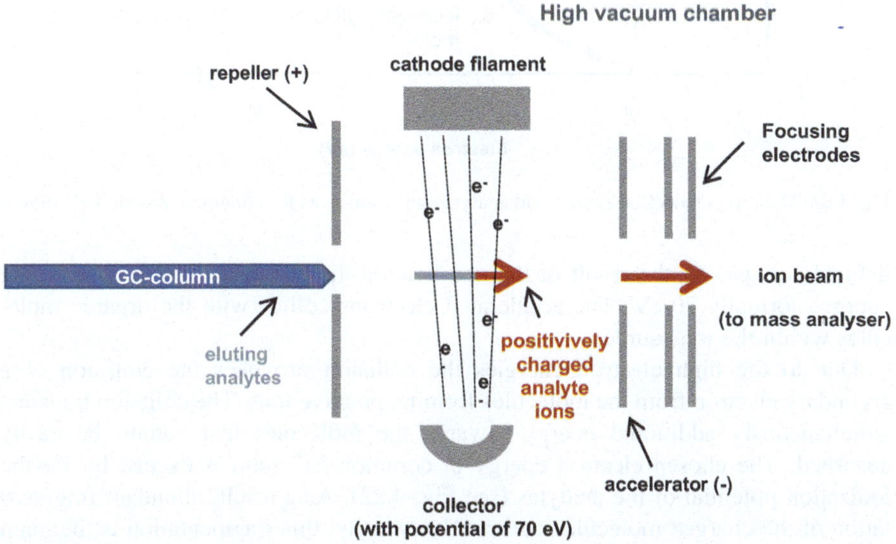

Fig. 4.21 Scheme of an EI$^+$ source, producing ions from the GC eluent and accelerate them into the mass analyser under high vacuum

precondition for a successful ionization. Once an ion is generated, it needs to become stabilized since ions are highly reactive. In order to avoid recombination of ions or secondary reactions the ions need to be kept in high vacuum, because under this condition the collision probability for the individual particles is minimized. Therefore, all mass spectrometers are equipped with high vacuum pumps producing low pressures around 10^{-5} to 10^{-6} bar. This high vacuum can be easily realized in GC/MS systems, since the overall gas flow into the ion source is low (around 1 mL/min). Therefore, ionization in GC/MS systems is performed directly in vacuo. Commonly, an electron impact ionization is used by a so-called EI$^+$ source (see Fig. 4.21). Here, electrons are generated by a cathode and become accelerated with a

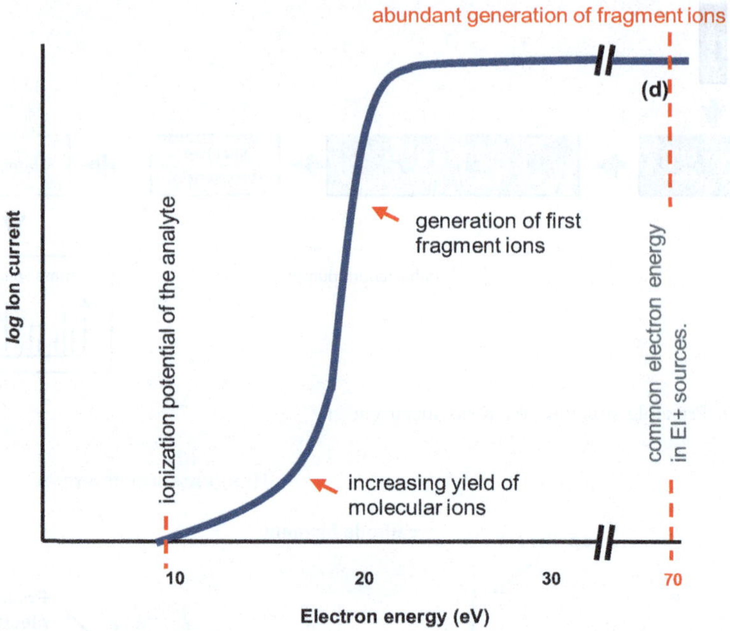

Fig. 4.22 Relation of electron energy and analyte ionization as well as fragmentation in EI^+ sources

defined voltage. As the result of the acceleration the electrons exhibit an accurate energy, normally 70 eV. The accelerated electrons collide with the organic molecules within the ion source.

Due to the high electron energies the collision provokes the emission of a secondary electron from the molecules forming positive ions. The collision transfers simultaneously additional energy towards the molecules that cannot be easily absorbed. The chosen electron energy in common EI^+ sources exceed by far the ionization potential of the analytes (see Fig. 4.22). As a result, abundant fragmentation of the charged molecules occurs. Noteworthy, this fragmentation is the main clue that allows to get information about the molecular structure of the analytes. After ionization the particles are accelerated and transported by focusing electrodes towards the mass analyzer.

Sometimes fragmentation is not requested but the measurement of intact molecules is needed. For this case, an alternative soft ionization can be used, the so-called chemical ionization CI. It uses the same type of ion source, but an additional gas is filled into the ion source chamber. The electron beam first ionized this gas and as a second reaction these ionized gas molecules collide with the analytes generating secondary ions. Since the energy transferred here is much lower, this ionization dominantly produces intact molecule ions. By using different auxiliary gases, the ionization energy can be tuned based on the different ionization potentials or proton affinity of the gases. Common gases are isobutane, ammonium or methane (see Table 4.2).

Table 4.2 Typical auxiliary gases used in CI sources

Auxiliary gas	Proton affinity (kJ/mol)
Methane	527
Isobutane	807
Ammonia	840
Triethylamine	966

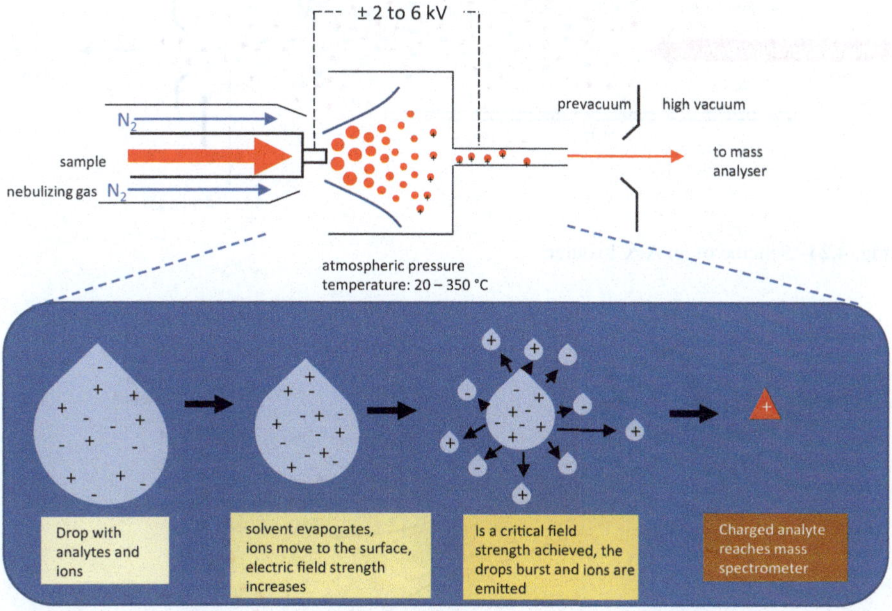

Fig. 4.23 Scheme of an ESI source and principals of ion formation in electrosprays

For LC/MS systems another ionization approach has been developed. Since high vacuum cannot be kept in systems with a continuous inflow of liquids around 0.5–1 mL/min (as common in liquid chromatography), the ionization in LC/MS systems occurs under normal pressure. Two methods are used, the electron spray ionization (ESI) and the atmospheric pressure chemical ionization (APCI).

As a first step, ESI sources produce an electrospray or aerosol by a nebulizer with a high voltage that is placed at the end of the liquid chromatograph (see Fig. 4.23). During the formation of the aerosol the solvent gets evaporated and the drops size decrease continously. At the end of the process ionized analytes are formed. This type of ionization is appropriate especially for macromolecular compounds and polar or even charged molecules. It is a soft ionization, hence fragmentation is not observed but the intact molecule ions are generated. Noteworthy, this kind of ionization produces also particles with multiple charges allowing to detect high molecular weight compounds. Higher charges reduce the m/z values to fit the normally measured m/z range.

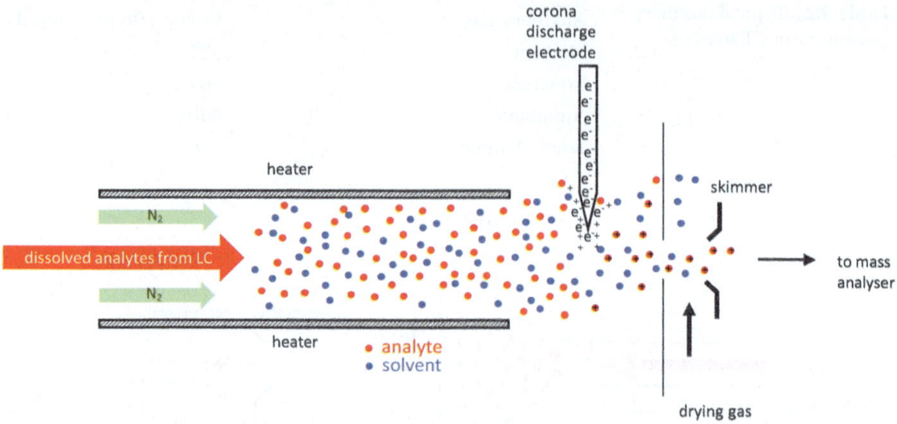

Fig. 4.24 Scheme of an APCI source

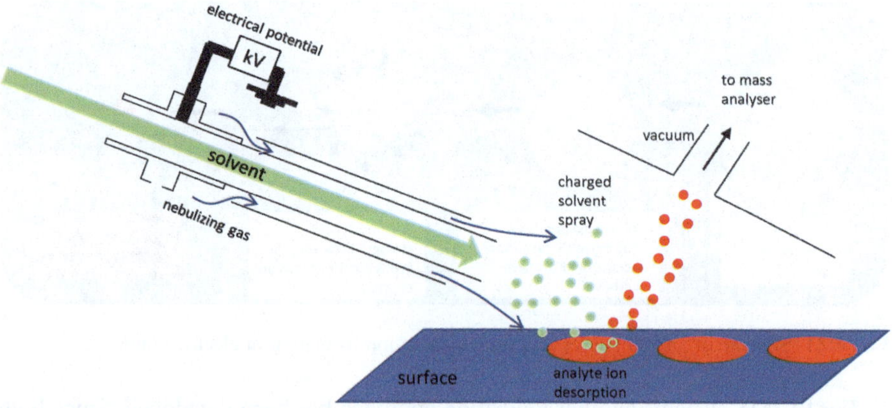

Fig. 4.25 Principal scheme of a DESI device

In APCI sources the ionization is supported by a corona discharge needle placed directly into the aerosol (see Fig. 4.24). First, mainly nitrogen and water molecules are transformed to ions in the aerosol. Subsequently, the analytes get ionized by these primary ions as a chemical ionization. Also this ionization is a soft one, comparable to ESI sources, but its application covers more low molecular and less polar analytes.

Instantly after ionization either by ESI or by APCI the ions are transferred into the mass spectrometer towards vacuum conditions preventing secondary reactions and recombination of ions.

A very special feature of ionization at normal pressure is realized in so-called desorption electrospray systems DESI (see Fig. 4.25). Here, the surface of samples gets directly ionized by an electrospray and subsequently analyzed by a mass spectrometer. Depending on the devices surface analyses can be performed in

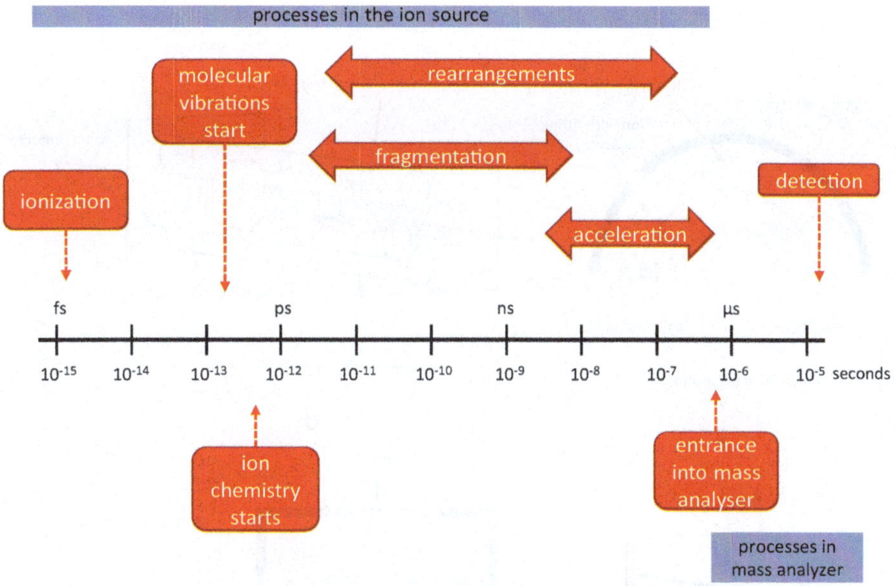

Fig. 4.26 Dimension of time for most relevant processes in mass spectrometry

different spatial resolutions. Applications of such technique are known e.g. for material analyses, geomaterials and pharmaceuticals.

Prior to discussing the different types of mass analyzer, a brief look on the dimension of time in mass spectrometric measurements needs to be done. As described, in ion sources the three principal processes ionization, fragmentation and acceleration (towards the mass analyzer) are performed. All three processes are fast but on different time scales as illustrated in Fig. 4.26. By far the most rapid process is ionization followed by fragmentation and rearrangements (in the range of fs to 100 ns). However, the overall residence time of the ions within the ion source is primarily determined by the acceleration lasting up to 1 μs. Noteworthy, this residence time is much lower as compared to the time that is needed for mass separation and detection of up to 0.1 ms. This mass analyzing time depends highly on the type of mass separation as discussed in the following.

As the second important unit in mass spectrometer, the mass analyzer separates the formed ions and detected them according to their m/z values or, in most cases, according simply to their mass. The first generation of mass spectrometer used magnetic sector fields in order to force the ions on circuits (see Fig. 4.27a). According to the Lorentz force, for all ions exhibiting the same impulse (as released from the ion source) the radius of the circuit is determined by m/z values as well as by the magnetic field force. The higher the mass (or m/z value) the longer the radius and vice versa, the lower the mass the shorter the radius. Secondly, the higher the magnetic field the shorter the radius for particles with same m/z values. Technically, the sectors in the sector field mass spectrometer allowed solely

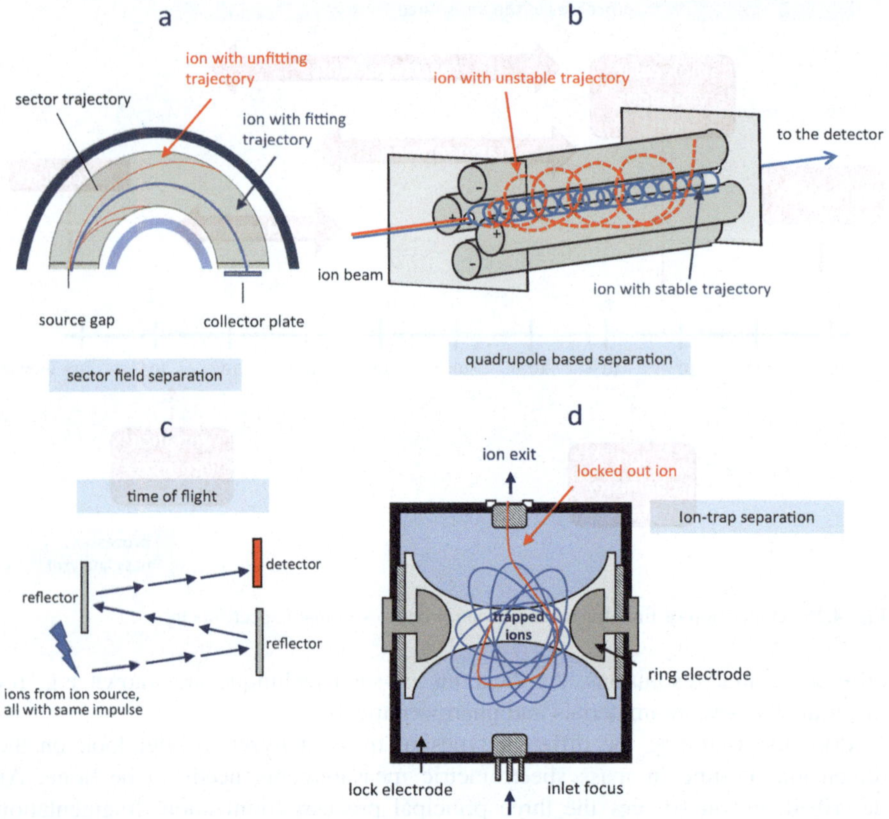

Fig. 4.27 Most common types of mass separators in mass spectrometry

one circular pass, that means at a constant and defined magnetic field strength only particles with one defined m/z ratio are able to pass the sector and to reach the detector. The measurement of a full mass spectra (commonly covering up to 1000 individual Dalton, Da, or unified mass units, u) is performed by rapid change of the magnetic field strength by jumping from one field strength to the next according to the masses. This type of measuring, also called scanning, detects ions for a very short time (around ms) at a given magnetic field strength corresponding to 1 m/z ratio, then it jumps to the next field value and starts the counting once again for the next m/z value. With this scanning approach most of the ions remain undetected, since all produced ions that do not fit the stable conditions for passing the sector at the given magnetic field strength get lost. This reduces the sensitivity of this technique but is still on a low level below ng in the described full scan mode. Further on, the time needed for one full mass spectrum is the sum of all individual measuring times of each individual mass and is commonly around 0.5–1 full scan per second. However, this relatively fast scanning allows a successful detection of GC separated peaks. An alternative method to enhance sensitivity is to detect only selected but specific ions of target analytes, that allows either a longer measuring time per mass, or a higher scan rate. This approach is named single ion monitoring SIM or single ion recording SIR and is dominantly used for sensitive quantitative analyses of preselected target analytes. Noteworthy, the full information on the compounds gets lost, since only very few ions or fragments are detected.

General Note
Fragmentation is a main clue to obtain structural information based on mass spectrometry. Gathering information on the smaller fragments allows insights into their molecular structure. Reassembling of all these moieties (like a puzzle) allows to characterize or even predict the structure of the whole molecule.

A significant improvement of mass accuracy and sensitivity has been developed early in the 1940s by adding an electrostatic sector between ion source and magnetic sector for focusing the ion beam prior to the mass separation. This double field mass spectrometers are still in use in many laboratories but are relatively expensive, since they can exhibit a high mass resolution (for mass resolution see Sect. 4.2.2).

Therefore, an alternative method has been developed. Instead of a simple magnetic sector field, a quadrupole field can be used (see Fig. 4.27b). The basic idea is the same, for ions with different m/z ratios stable and instable trajectory exist also in quadrupole fields allowing a scanning similar to the sector field mass spectrometer. Core of the separator are four metal tubes fed by both DC and AC voltage resulting in a quadrupole field that is tunable by the individual voltages. The quadrupole field forces the ions on a sort of coil pathway. Under stable conditions the radius remain constant, but for lower masses the radius increase with time and pathway, whereas for higher masses the radius decreases. Both the sensitivity as well as the scan rate are similar as compared to sector field mass spectrometer, but the mass resolution is low.

A further technical approach in mass spectrometry is based on a linear acceleration of the ions. Once all ions get the same impulse at a defined time point their velocity is different according to their m/z ratio and, consequently, their time of flight for a certain distance. This phenomenon is used in time-of-flight mass spectrometry (see Fig. 4.27c). The time the ions need for a defined pathway is measured and the time is directly correlated with the m/z values. Advantage of this technique is a contemporary detection of all generated ions and a resulting higher sensitivity. Further on, the mass resolution can be very high depending *inter alia* on the flight distance. This points to one challenge of this technique to build devices with sufficient stretches of way. In modern devices the pathway is optimized by reflecting the ion beam to increase the distance.

Finally, a fourth approach is used in mass separation that also detects all ions generated. The name of this technique exactly describes the method, the ion-trap mass spectrometry (see Fig. 4.27d). The ions get trapped in a special trap consisting of different electrodes producing dynamic electrostatic fields. The resulting field directs the ions on complex but stable orbits within the ion source. Under these

Table 4.3 Brief comparison of most common mass spectrometers

MS type	Mass resolution	Sensitivity	Costs
Sector field	High	Low	High
Quadrupole	Low	Low	Low
Ion-trap	Low–high	High	Low–high
Time-of-flight	High	High	High

conditions all ions can be stored for a prolonged time period. Following individual ions can be forced by changing the field conditions to leave the ion-trap for a subsequent detection. These mass spectrometers are also very sensitive and exhibit higher mass resolution.

Technical improvement of this basic principle is achieved by the ion cyclotron resonance mass spectrometry, that needs Fourier-transformation (FT-ICR MS), and the so-called Orbitrap. Both are devices with very high sensitivity and high mass resolution.

All different types of mass spectrometer exhibit their own advantages and disadvantages and are therefore techniques commonly used in different applications and laboratories. A very rough differentiation can be done taking the three basic parameters sensitivity, mass resolution and costs into account (see Table 4.3).

As one can see, mass resolution is an important parameter in mass spectrometry. In simple word, it is the accuracy in mass detection and can be illustrated as signal width of the individual masses detected. Considering a given signal width, two neighbored signals need to have a minimum distance to become resolved as illustrated in Fig. 4.28. Or with other words, if the signals are nearby and have a broad width, they appear as one superimposed but not well separated signal. Formally, the resolution of a mass spectrometer is given relative to the measured masses. A low resolution of around 1000 means that a mass is accurate by 1/1000 of the measured value. For a mass of 500 Da this implies an accuracy of 0.5 Da or, simply spoken, a second signal with a difference of 0.5 Da can be detected just separately. High resolution mass spectrometry allows mass resolution of 5000 up to 1,000,000.

The higher the mass resolution, the better the separation of similar but not identical masses. This issue becomes interesting by a closer look on the atomic masses. Ignoring decimals leads to the so-called nominal masses, e.g. 12 Da for carbon, 1 Da for hydrogen or 16 Da for oxygen. If we have a look on fragments and molecular ions in mass spectra that means e.g. 15 Da for a methyl group or 32 Da for methanol. However, atom masses are not nominal, but have variances as exemplified for some elements in Table 4.4.

These variances have implication for the exact masses of fragment and molecular ions measured in mass spectrometry. The exact mass of a methyl group is not 15 Da but 15.023475 Da. Further examples of nominal vs. exact masses are given for the molecular masses of some selected pollutants in Table 4.5.

If a mass spectrometer is able to measure masses very accurately, it enables to resolve or deconvolute signals with the same nominal masses but different molecular compositions. This is the main clue of exact mass spectrometry: to obtain information on the elemental composition of measured masses (fragments or molecular ions,

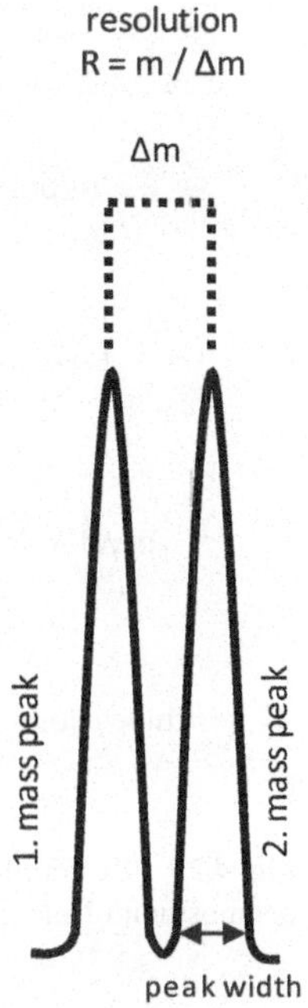

Fig. 4.28 Illustration of mass resolution

Table 4.4 Exact masses of elements highly relevant in organic chemistry

Element	Isotope	Dalton (Da)
Hydrogen	^{1}H	1.0078250
Carbon	^{12}C	12.000000
Nitrogen	15N	14.003074
Oxygen	^{16}O	15.994914
Chlorine	^{35}Cl	34.968854

Table 4.5 Nominal and exact masses of selected organic pollutants

Compound	Exact mass	Nominal mass
Hexachlorbenzol ($^{12}C_6{}^{35}Cl_6$)	281.81312	282
Carbamazepin ($^{12}C_{15}{}^{1}H_{12}{}^{14}N_2{}^{16}O$)	236.09496	236
2,3,7,8-Tetrachlordibenzo-*p*-dioxin ($^{12}C_{12}{}^{1}H_4{}^{35}Cl_4{}^{16}O_2$)	319.89655	320

as illustrated in Fig. 4.29) or to use the exact composition for very sensitive detection. However, most measurements in Organic Geochemistry are performed as low-resolution mass spectrometry.

Finally, one modern development in mass spectrometry needs to be addressed, the linkage of several mass spectrometer units in one device also known as multidimensional mass spectrometry. Most common is the MSxMS technique (also called tandem MS) linking two mass spectrometric units or mass analyzer, e.g. two quadrupol analyzer or a quadrupol-ToF combination (see Fig. 4.30). The clue is related to the ionization methods. Normally a first analyzer uses soft

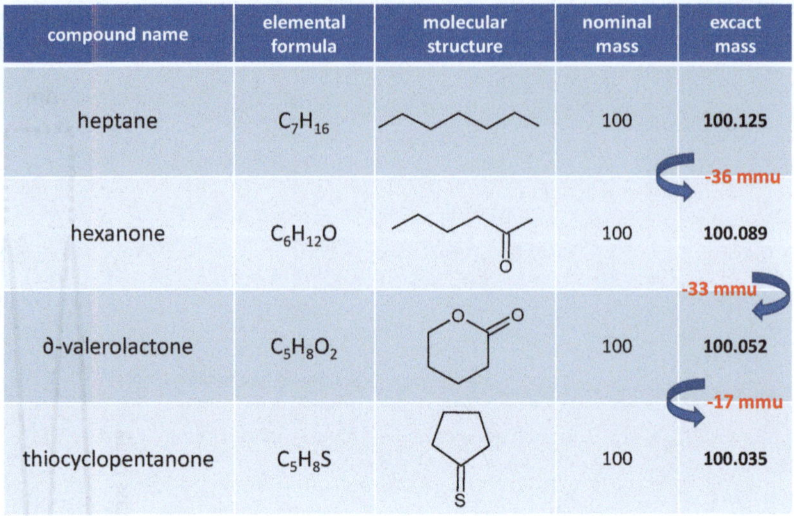

Fig. 4.29 Illustration on how exact mass spectrometry enables to obtain information on elemental composition of measured signals

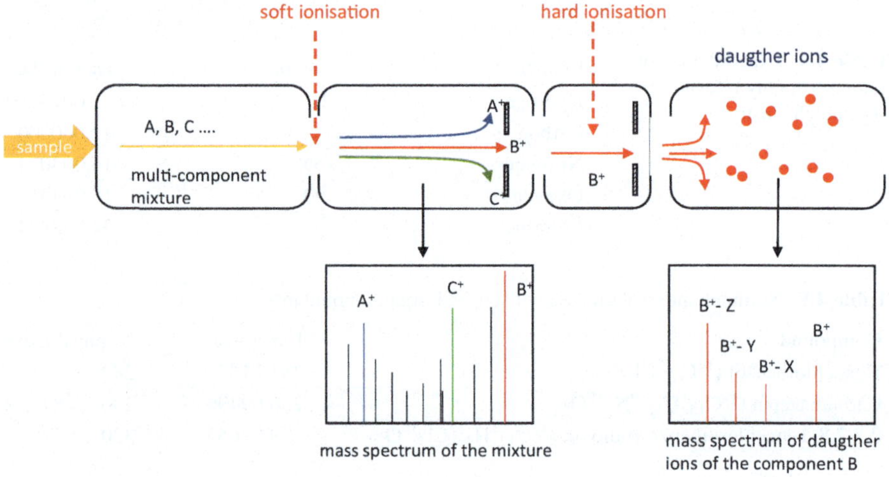

Fig. 4.30 Scheme of a MSxMS device and the type of measurement (modified after Cammann 2010)

ionization in order to obtain intact molecular ions, that can be separated. Then selected ions can be transferred to the second mass analyzer for a second measurement. Prior to this second separation a hard ionization induces fragmention forming fragment ions (so-called daughter ions) from the isolated molecular ion. This process enables a first isolation of a substance by the molecular ion and, subsequently, the detection of structural information by corresponding fragments. This approach enables a specific detection of individual compounds also in mixtures that cannot be fully separated by chromatography. Furthermore, the sensitivity of such analyses is commonly high.

4.2.2 Mass Spectra

As already described, mass spectrometry measures the intensity of formed ion to mass ratios (m/z). The standard techniques in MS generate monocharged ions, hence the m/z is normally restricted to mass (with only some very rare exceptions, e.g. double charge aromatic moieties). Therefore, a mass spectrum consists of an x-axis appointed to m/z values and a y-axis with intensity. The intensity is not given as absolute values but as relative ones, normalized to the highest signal with 100%. An example is given in Fig. 4.31. Signals are entitled as 'peaks' and some are very specific. The peak with the highest intensity (the 100% peak) is commonly named *base peak*, whereas the signal representing the intact molecule is called *molecular peak*. Further on, for several elements not only one mass but also isotope masses (e.g. $^{12}C/^{13}C$ or $^{35}Cl/^{37}Cl$) exist, which produce certainly also signals in mass spectra, named *isotope peaks*.

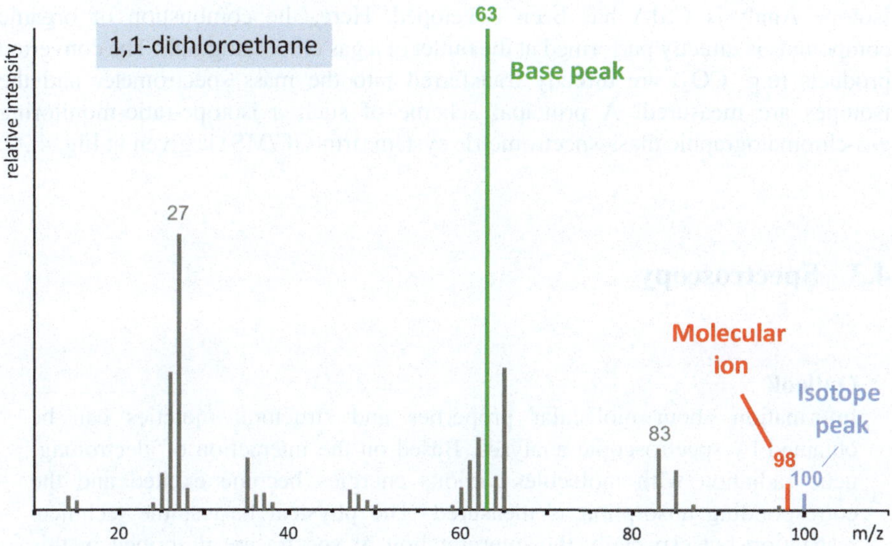

Fig. 4.31 Scheme of a typical mass spectrum

4.2.3 Stable Isotope Mass Spectrometry

A special type of mass spectrometry has been applied intensively in Organic Geochemistry, that is neither focused on getting structural information on the compounds nor is used for quantification. This approach aims at measuring the isotope composition of the organic matter or even of individual compounds, the so-called stable isotope mass spectrometry.

In Organic Geochemistry a general interest exists to detect the relative composition of stable isotopes of carbon and to a minor extent of hydrogen, oxygen and nitrogen. The isotope composition is influenced by various geo- and biochemical processes as well as environmental conditions. As the main clue, for detection of the isotopes the samples are converted into one representative species via combustion or pyrolysis. For stable carbon isotope analyses an exhaustive combustion of the organic matter produces CO_2 as analyte for further mass spectrometric measurements. The mass spectrometer just measures the amount of $^{12}CO_2$ with 44 m/z and $^{13}CO_2$ with 45 m/z. From these quantitative values the carbon isotope ratio ($\delta^{13}C$) can be easily calculated. The same approach analyses organically bound nitrogen and hydrogen isotopes by conversion into N_2 and H_2, respectively, and measuring of H_2/HD or of the ^{14}N and ^{15}N signals.

This approach is applied to complete samples revealing insights into the isotope composition of bulk material. However, it is also of high interest to get information about the isotope composition of individual compounds (e.g. *n*-alkanes, isoprenoid biomarker, BTEX pollutants). For this purpose, the so-called Compound Specific Isotope Analysis CSIA has been developed. Here, the combustion of organic compounds is directly performed at the outlet of a gas chromatograph. The converted products (e.g. CO_2) are directly transferred into the mass spectrometer and the isotopes are measured. A principal scheme of such a isotope-ratio-monitoring gas-chromatographic mass-spectrometric system (irm-GC/MS) is given in Fig. 4.32.

4.3 Spectroscopy

Outlook
Information about molecular properties and structural moieties can be obtained by spectroscopic analyses. Based on the interaction of electromagnetic radiation with molecules various energies become excited and the corresponding absorption is measured. The physical basics, the technical realization but especially the interpretation of spectra are described in this chapter.

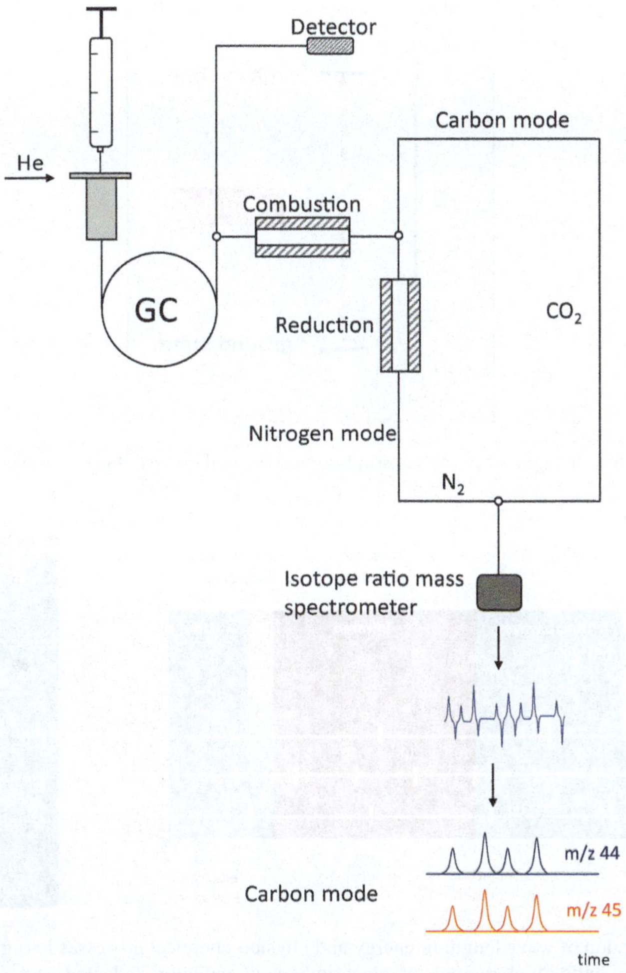

Fig. 4.32 A principal set-up of an irm-GC/MS system for compound specific stable carbon or stable nitrogen isotope measurements

4.3.1 Principles of Spectroscopy

Spectroscopy is used as very valuable tool to gain insights into the structural properties of organic molecules. The method is based on a very simple physical phenomenon, the interaction of electromagnetic radiation and matter, also known as 'absorption' and its counterpart 'emission'. Since this interaction occur on molecular or atomic dimensions, it can be explained only by quantum mechanics. For this purpose, electromagnetic radiation can be described either as particle beam or as a wave, the so-called wave-particle dualism. For describing the spectroscopic processes, the most important aspect is the quantization implying only discrete energy

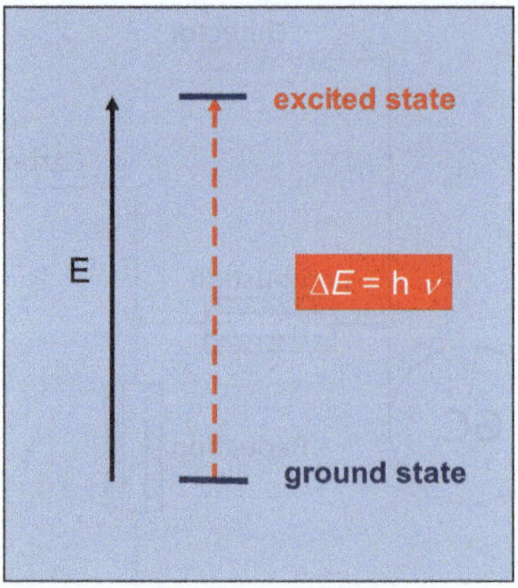

Fig. 4.33 Basic principles of photon absorption and excited energy states in molecules

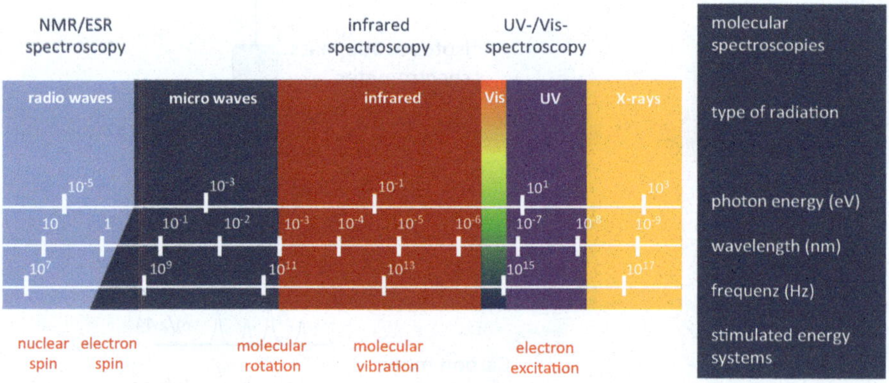

Fig. 4.34 Relation of wave length or energy and physico-chemical processes in organic molecules affected by the individual regions of electromagnetic radiation (adapted and simplified after Schwedt 2007)

states in molecular systems. Hence, changing any energy state in molecules is only possible by adding a corresponding energy package. With respect to absorption processes and describing electromagnetic radiation as particles, the absorption is only possible if the energy of the absorbed photon fits exactly the energy difference between initial or ground state and excited state (as exemplified in Fig. 4.33). Since many energy states and their distances depend on the molecular structures and atomic properties, the absorption energy can provide information on molecule and atom characteristics. This is the basic approach of analytical spectroscopy.

The variation of spectroscopic methods is closely related to the broad spectrum of wave lengths of electromagnetic radiation (see Fig. 4.34). Wave lengths from meter down to nanometer scale reflect a wide scope of corresponding energies. Therefore, the individual energy states excitable by photon absorption are comprising various mode of induced physico-chemical changes in molecules. Low energy radiation interacts with the spin energies either of atomic nucleus or electrons forming the base for NMR and ESR spectroscopy. Radiation with somewhat higher energy provokes changes in the rotation and vibration of molecules. In particular the later one is related to the wave length area of infrared radiation and, therefore, this area is used for infrared or IR spectroscopy. In the energy range of visible and ultraviolet light, molecules are able to absorb the photons by changing the energy state of their valence electrons leading to the application of UV/Vis spectroscopy. At higher energies in the X-ray range the inner electrons become excited to release the atom representing the base for atom absorption spectroscopy AAS.

However, in contrast to NMR/ESR, IR and UV/Vis spectroscopy applied (not exclusively but intensively) for the analysis of organic molecules, AAS is a method for elemental analysis and not common in organic analyses. Further on, NMR spectroscopy as an extraordinary powerful tool for structure elucidation has only a very limited application in Organic Environmental and Geochemistry due to some aspects: (i) this method needs a relative high amount of material or analytes in the range of mg, (ii) information on chemical structures of substances are only available for isolated material (no complex mixtures), and (iii) the time of measurement is high (minutes to hours) and, consequently NMR is not successfully linkable to chromatographic techniques such as gas or liquid chromatography. These restrictions avoid the analyses of environmental or fossil samples that are characterized by complex mixtures at partly very low concentrations, that normally need an intensive separation by chromatography. However, some special applications of NMR in Organic Geochemistry exist, e.g. the characterization of humic substances in soil by solid state NMR.

Consequently, the following two chapters focus on the UV/Vis and IR spectroscopy.

4.3.2 UV/Vis Spectroscopy

As the basic physical principle of UV/Vis spectroscopy, the energy of UV/Vis radiation fits the energy differences of the valence orbitals in organic molecules. In simple words, the absorption of a photon in the UV/Vis range provokes a valence electron to jump from its ground state orbital to an orbital with higher energy. Energies of molecule orbitals are certainly quantized and, therefore, only selected differences are possible leading to clear restricted transition. A general differentiation of orbitals available for the transition are given in Fig. 4.35.

In ground state the valence electrons are either in bonding σ- or π-orbitals as well as non-bonding orbitals. From these orbitals they can jump into anti-bonding σ^*- or π^*-orbitals. It is obvious, that transitions from σ-orbitals to higher energy orbitals need more energy as compared to π-orbitals or non-bonding orbitals. Hence, electrons in double bounds (π-orbitals) or from free electron pairs (e.g. nitrogen, oxygen) can be used for absorption at lower energy, whereas electrons in single bonds need higher energy for getting excited. Further on, the chemical setting influences the energy levels of the individual molecular orbitals and, thus, the resulting energy

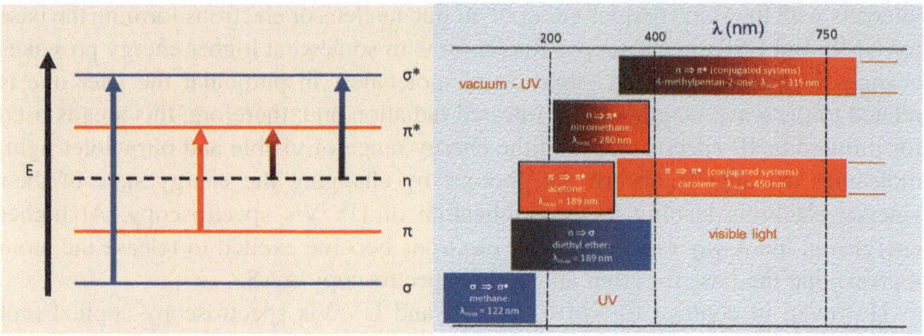

Fig. 4.35 Orbital energy levels and electron transitions used for UV/Vis spectroscopy (left). The corresponding absorption regions for the different transitions with some examples (right); (partly simplified and modified after Cammann 2010)

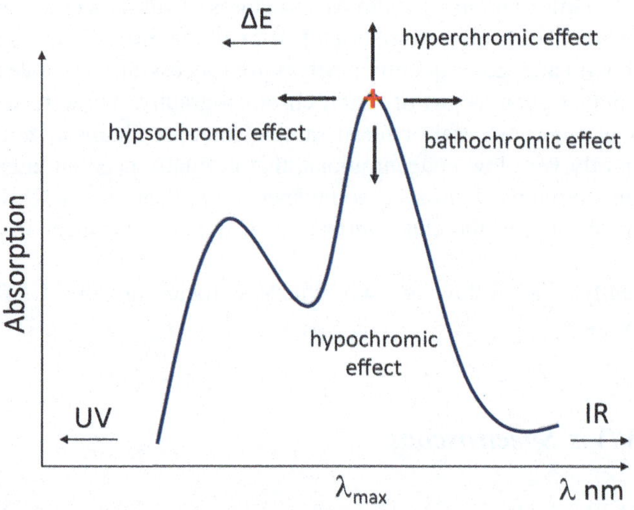

Fig. 4.36 Principle shifts of absorption and intensity resulting from different chemical properties

differences. Variations in chemical structure can result in a shift towards higher or lower absorption energy, the so-called hypso- and bathochromic effects, or to reduced or enhanced intensity, the hyper- and hypochromic effect (see Fig. 4.36).

As a main consequence, the absorption energies reflect the chemical structure of the analyzed molecules. Or vice versa, the chemical composition induces different absorption wavelengths. This is the reason why UV/Vis spectra differ for different substances and can therefore be used for getting information about the investigated molecules. The energy regions and corresponding electron transitions correlated with structural moieties or exemplifying compounds are given in Fig. 4.35.

As a general order, the absorption maxima shifts from high energy to lower energy from C-C or C-H single bonds over bonds with oxygen or nitrogen atoms to double bonds (C=C, C=O ...). For absorption in the region of visible light structural moieties like conjugated double bonds are needed, since the conjugation lowers the relative energy differences between the π- and π*-orbitals down to nm, the visible light. Following this argument, it becomes clear why organic molecules like carotene or chlorophyll, with their extended conjugated double bond systems, are color pigments.

Excursus: ***How UV/Vis absorption can explain bond cleavage***
Since valence orbitals are responsible for the bonding between atoms, UV/Vis absorption has an influence on these bindings. If an UV/Vis absorption shifts an electron from a bonding orbital to a non-bonding orbital the bond losses overall bonding energy. The energetic advantage of forming molecular orbitals from atom orbitals (the formal description of forming molecular bonds) is gone, the energy balance of an excited bond does not support binding, hence a cleavage might be observed over time.

The most common technical realization of UV/Vis spectroscopy follows simple approaches. Normally, UV/Vis spectrometry is applied to solutions of analytes. One very common spectrometer works with a double beam method. From a polychromatic source, a selected wavelength (isolated by a monochromator unit) irradiates both a reference cell and the sample. Thereafter, the intensity of both beams is compared, and the measured difference represents the absorption. Passing through all wavelength and recording the corresponding absorption (or extinction) enables to build up the corresponding spectrum. On the contrary, a single beam approach irradiates the sample with the entire spectrum of wavelength simultaneously and the resulting residual radiation is separated by a diffraction grating and all wavelength are measured also simultaneously in an array detector (Fig. 4.37).

The result of an UV/Vis analyses is the corresponding spectrum, normally a correlation of absorbance with wavelengths. An example is given in Fig. 4.38. UV/Vis spectra are characterized by very broad absorption signals or even regions. Generally, an attribution of absorption regions to chromophoric moieties is possible.

But although the spectra are characteristic for substances, UV/Vis spectroscopy is not common for qualitative analyses due to the very low specificity of this approach. There is another field of application, in which UV/Vis spectroscopy is more often used. Detecting only one characteristic wavelength allows, according to the Lambert-Beer Law (see Fig. 4.39), a direct linear correlation of absorption intensity and amount or concentration. Hence, UV/Vis spectrometer are widely used for quantitative determination of organic compounds and, fortunately, are easily linkable to liquid chromatography. As an example, aromatic compounds (e.g. PAHs) are often determined by UV/Vis spectroscopy using the absorption area of the aromatic π-electrons at around 254 nm. Applying this method on soil extracts by LC-UV/Vis allows an easy and fast determination of these pollutants.

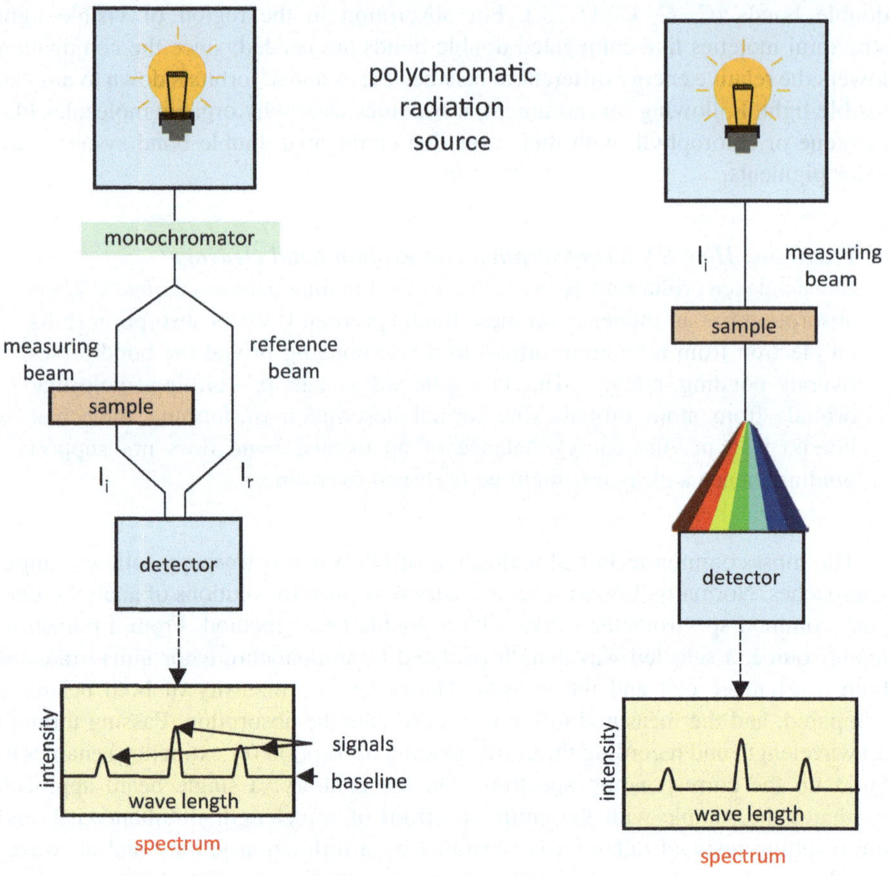

Fig. 4.37 Scheme of an UV/Vis spectrometer

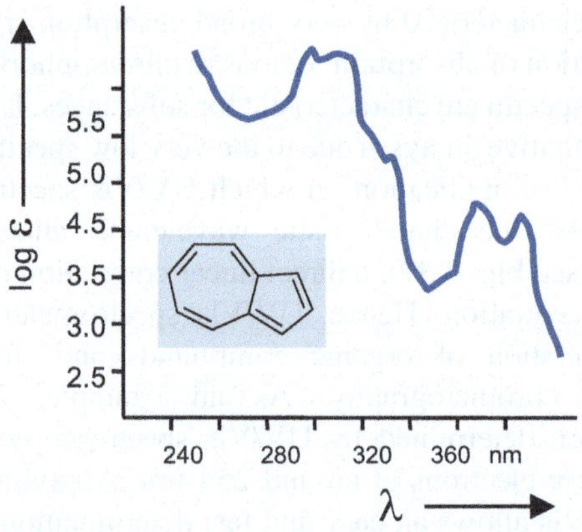

Fig. 4.38 UV/Vis spectrum of azulene

Lambert-Beer Law

$$\text{Extinction } E_\lambda = \log \frac{I_0}{I_d} = \varepsilon \cdot d \cdot c$$

I = Intensity (before I_0 and after I_d absorption)
ε = molar attenuation coefficient
c = concentration of absorbent
d = pathlength

Fig. 4.39 The Lambert-Beer Law pointing to a linear correlation of concentration and extinction/absorption

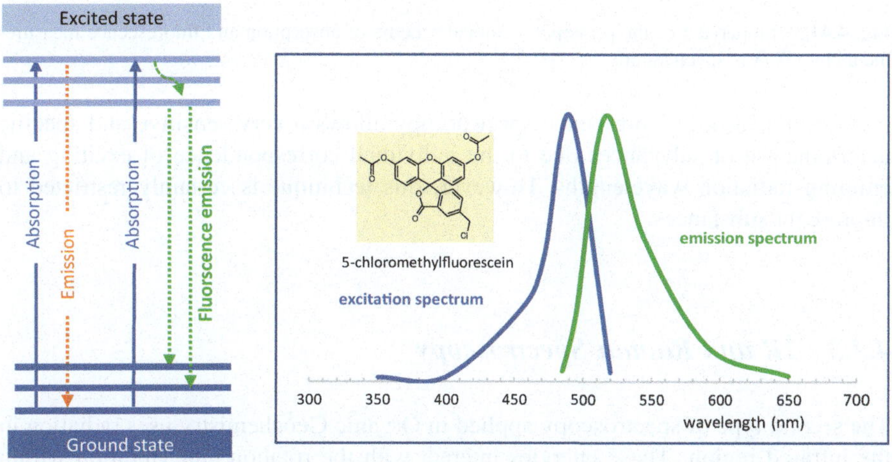

Fig. 4.40 The energetic principles of the fluorescence phenomenon (partly simplified and modified after Cammann 2010)

For a more accurate and more sensitive determination an alternative analytical approach is used by UV/Vis spectroscopy. It benefits from the phenomenon of fluorescence. For some molecular systems no direct accordance of adsorbed and emitted wavelength has been observed. Normally, the reversed effect of absorption is the emission of radiation with the same energy as compared to the absorbed one. In fluorescence, in the excited state there is not a direct emission but a transfer on a slightly lower energy level. The following final stabilization to the ground state causes an emission radiation with somewhat lower energy as compared to the absorbed one (see Fig. 4.40).

As a consequence, a fluorescent substance can be excited with radiation at higher energy but emits radiation with lower energy. Fluorescence spectroscopy follows this approach by exciting the target analytes but to measure the fluorescence (see Fig. 4.40). A technical realization with a 90° angle detection of the emitted radiation is given in Fig. 4.41.

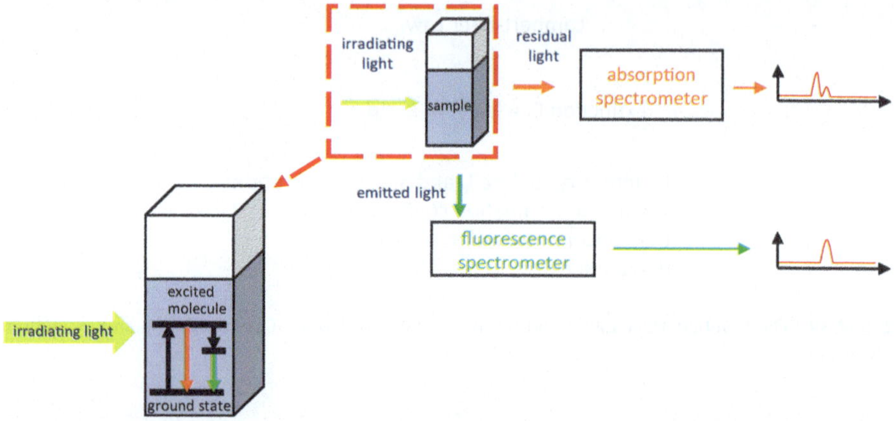

Fig. 4.41 Comparison of the principal technical scheme of absorption and fluorescence measurements by UV/Vis spectroscopy

In principal, the fluorescence spectroscopy allows a very sensitive and specific determination of substances due to the individual correspondence of exciting and emitting radiation wavelengths. However, this technique is certainly restricted to fluorescent substances.

4.3.3 IR and Raman Spectroscopy

The second type of spectroscopy applied in Organic Geochemistry uses radiation in the infrared region. These energies interact with the rotation and vibration within molecules. The corresponding energy states can be described by the asymmetric potential well of an anharmonic oscillator (see Fig. 4.42). All vibrations and rotations in an organic molecule are quantized, therefore, IR absorption represents the transition from a lower rotation/vibration level to a higher one. Noteworthy, not all organic substances are IR active. There are some prerequisites that need to be fulfilled to have an IR absorption. In particular there must be a change of the dipole moment of the molecule during the vibration. As a consequence, some substances remain IR inactive.

Excursus: ***How IR absorption can explain bond cleavage***
Looking on the potential energy of vibrations a little closer, two aspects become obvious. Firstly, there is no energy state at zero energy or energy minimum. That means, also at the lowest energy (*absolute zero point*) there is a minimum of vibration, or with other words, a molecule cannot be frozen to

(continued)

absolute standstill. Secondly, the asymmetric shape of the potential well implies that the differences at the very high energies become smaller and smaller. Concurrently, the average distance from the bonding partner increases rapidly. This can be interpreted as bond cleavage as result of IR energy absorption. At high energies the bond distance is infinite, hence the atoms are cleaved. Therefore, the asymptotic maximum at the right axis of the potential well correlates with the dissociation energy.

Technically, IR spectroscopy is realized by two types of spectrometer. First, a double beam spectrometer is constructed comparable to UV/Vis spectrometer as described in Sect. 4.3.2 and illustrated in Fig. 4.38. To overcome the relatively long measuring time as well as the uncomfortable successive screening of individual wavelengths, more often Fourier Transformation IR spectrometer (FTIR) are used. Here, the sample is irradiated by the full band of wavelengths, and the superimposed absorption signals become deconvolved by a Fourier transformation using an overlaid self-interfering laser beam (see Fig. 4.43). This technical realization allows a much faster measurement with a partly higher sensitivity.

Noteworthy, IR spectroscopy can be applied to liquids (including solutions of analytes), solids and gases. For these purposes, different sampling holder and devices have been developed (see Fig. 4.44). As a common feature, all materials, which are irradiated but do not represent the samples, consist of IR inactive substances such as potassium bromide KBr or sodium chloride NaCl. As an example, solids are commonly measured by incorporation the sample material into a KBr

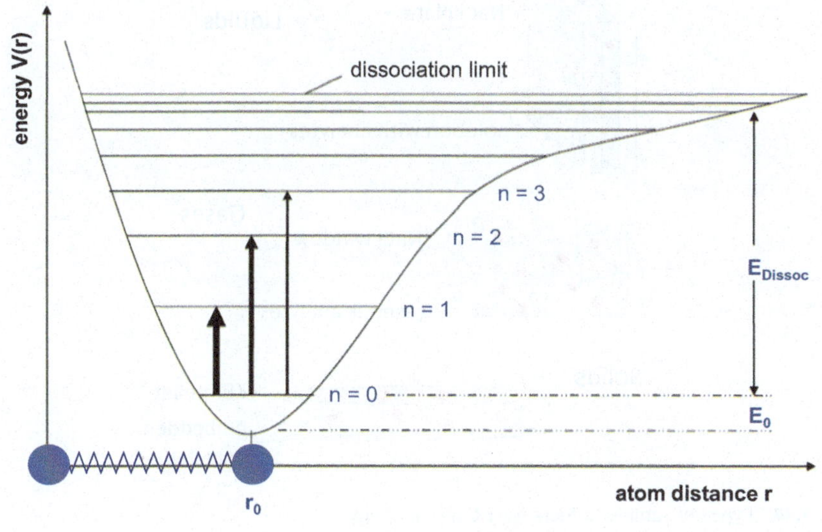

Fig. 4.42 The potential of an anharmonic oscillator as description for molecular vibration energy states

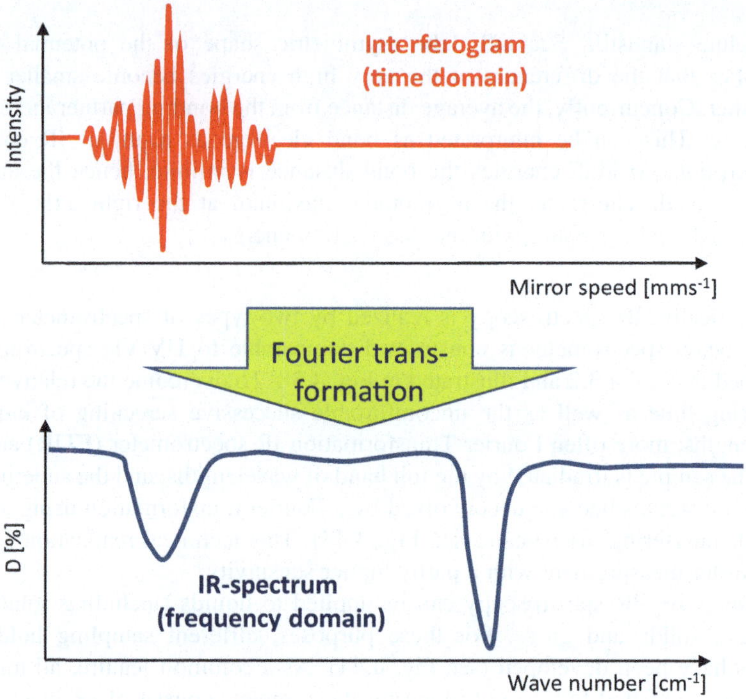

Fig. 4.43 Scheme of Fourier Transformation in FTIR

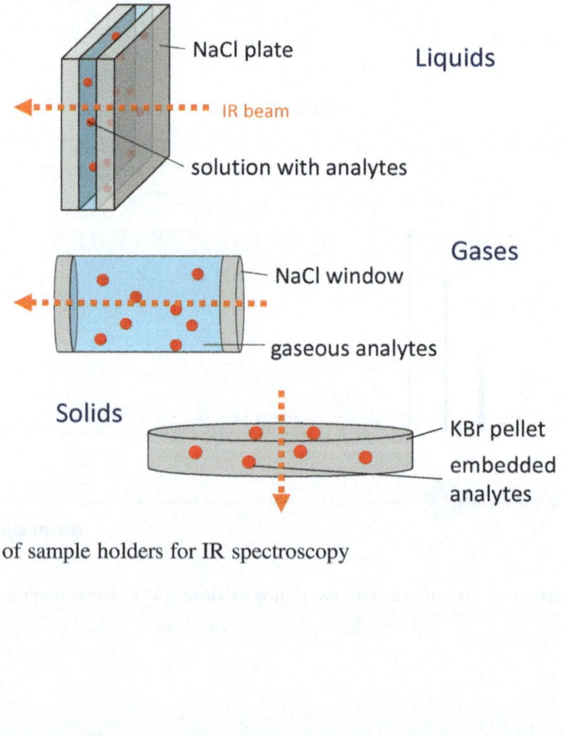

Fig. 4.44 Types of sample holders for IR spectroscopy

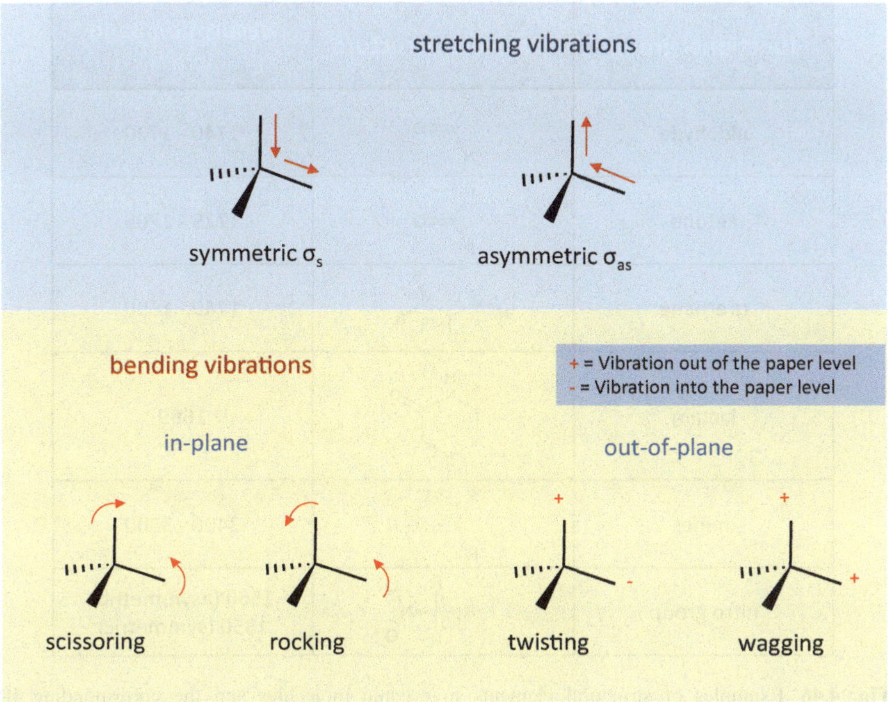

Fig. 4.45 Classification of molecular vibrations

pellet. Compression of this pellet forms a transparent matrix of crystalline KBr in which the solid sample is fixed. This pellet can be directly measured in an IR beam without interferences by the KBr matrix.

For common IR spectroscopy the vibration transitions are used. The corresponding vibrations within a molecule can be divided into different groups of vibrations, which are summarized in Fig. 4.45. A first group of vibrations are along the bonding axis, the so-called valence vibrations. They are subdivided into symmetric and asymmetric ones. On the contrary a larger group of vibration types are characterized by changing bonding angles forming the so-called deformation vibrations. These vibrations can be differentiated by their oscillation plane into in-plane and out-of-plane types. Having a closer look on the type of deformation, a former partition allows to distinguish bending and rocking (in-plane) from twist and wagging (out-of-plane) vibrations.

In a complex molecule all these types of vibrations along all bonds are realized contemporarily. Frequencies and magnitudes of the vibrations depend on the atoms involved, in particular their masses, their chemical as well as steric properties. This linkage of structural properties with vibrations and finally with absorption wavelengths allow to deduce molecular information from IR spectra.

Commonly, for IR spectra not wavelengths but their inverse value, the wavenumber ν (in cm^{-1}), that is directly correlated with absorption energy.

functional group	molecular structure	absorption band [cm^{-1}]
aldehyde	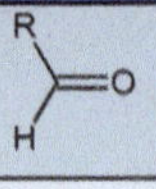	1740 - 1720
ketone	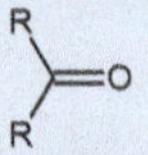	1725 - 1705
urethane	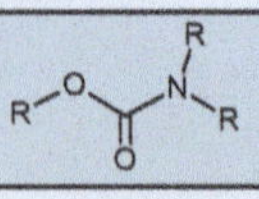	1740 - 1690
lactam	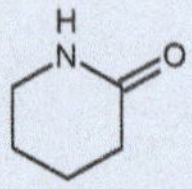	1669
imines	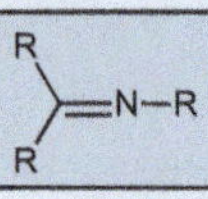	3400 - 3300
nitro group	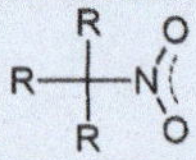	1560 (asymmetric) 1350 (symmetric)

Fig. 4.46 Examples of structural elements in organic molecules and the corresponding IR absorptions

Unfortunately, the IR spectra are normally reported with an x-axis covering wavenumbers between approx. 400–4000 nm^{-1} but with decreasing values. Further on, not the absorption strength is given on the y-axis but with the transmission leading to a negative excursion of the measured values in case of absorption. An example is given in Fig. 4.46.

However, the interpretation of IR absorption in a spectrum seems to be difficult due to the complex vibrations and the numerous parameters influencing them. Fortunately, as a result of complex theoretical considerations, a very simple interpretation approaches allows the direct linkage of functional groups and isolated structural elements to observed absorption bands in IR spectra. With simple words, for a given structural moiety (e.g. an aliphatic double bond, a nitrile group or an aldehyde group) a clearly defined absorption area can be found. For some functional groups and structural elements these absorptions are summarized in Fig. 4.46 demonstrating the accuracy and the detailed structural information that can be gained from IR spectra.

For a systematic ab initio interpretation of IR spectra, a simple approach using a direct linkage of absorption bands or regions and structural moieties is applicable. These correlations are illustrated in Fig. 4.47. Briefly, five different regions can be differentiated. From 3800 to 3200 cm^{-1} OH- and NH- bands with medium absorption intensity are visible. The broad shape of these signals is related to the formation

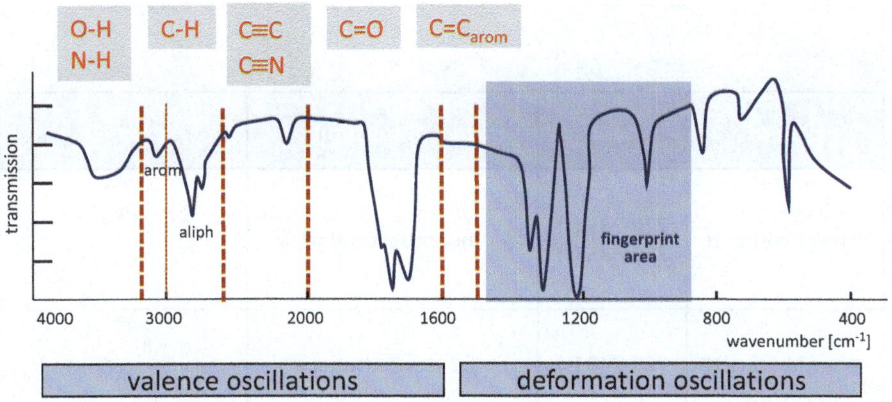

Fig. 4.47 Systematic allocation of absorption regions and main functional groups or structural elements

of hydrogen bonds and the corresponding width of intramolecular interactions influencing the vibration energies. Noteworthy, the OH-bonds in carboxylic acids appear also as broad signals but at lower ranges (3200–2500 cm^{-1}). The next very interesting region around 3000 cm^{-1} covers the CH-bands. A detailed look allows to separate aromatic/unsaturated aliphatic (bands above 3000 cm^{-1}) from saturated aliphatic CH bonds (bands below 3000 cm^{-1}). The intensities of both types of CH bonds differ significantly. The aromatic bands have a lower intensity as compared to the aliphatic ones. Following, the region between approx. 2800 and 2000 cm^{-1} covers absorptions of triple bonds either between carbon atoms (alkines) or carbon and nitrogen atoms (nitrile groups). Between 2000 and 1600 cm^{-1} very intensive and prominent signals are related to the carbonyl groups (ketones, aldehydes, carboxylic acids, chinones etc.). In the range between 1600 and 1500 cm^{-1} two or three signals appear for aromatic C-C bonds. Position and intensity vary. Then, a very specific region between 1500 and approx. 1000 cm^{-1} has been described as fingerprint region. In this range various different types of molecular moieties exhibit absorption bands of deformation vibrations with highly varying intensities. This superimposition does not allow a distinct interpretation but forms a substance specific absorption pattern that led to the denomination ‘fingerprint area’. Wavenumbers below 1000 cm^{-1} can be attributed to more specific absorptions, e.g. bonds between carbon and halogen atoms. Further on, a systematic and thorough analysis of the region between 950 and 700 nm^{-1} allows sometimes the characterization of substitution pattern at aromatic systems as exemplified in Fig. 4.48.

As already mentioned, some substances do not exhibit appropriate IR activity, they are ‘invisible’ for IR spectroscopy. However, there is a complementary method for analyzing vibration energies that allows to measure also such compounds, the Raman spectroscopy. Here, the molecule gets excited by a primary monochromatic Laser radiation with higher energy in the range of visible, near infrared or near UV radiation. Interestingly, slight differences between absorbed and subsequently

substitution pattern of an aromatic ring			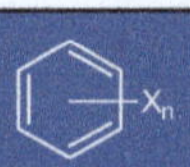
pattern	**band**	**degree of substitution**	**example**
5 neighboring H	770-735 (s) 710-685 (s)	monosubstitution	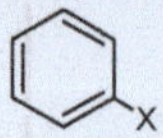
4 neighboring H	760-740 (s)	1,2-disubstitution	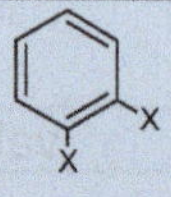
3 neighboring H	800-770 (s)	1,3-disubstitution 1,2,3-trisubstitution	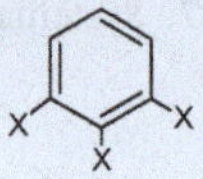
2 neighboring H	840-800 (s)	1,4-disubstitution 1,3,4-trisubstitution ...	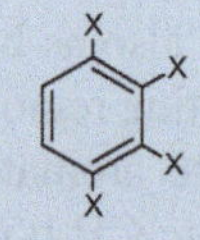
Isolated H	900-800 (w)	1,3-disubstitution 1,3,5-trisubstituion ...	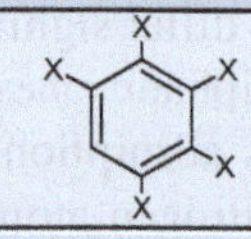

Note: multiple pattern are possible,
e.g. 1,3-substitution implies both an *isolated H* and *3 neighboring H*

Fig. 4.48 IR absorptions for interpretation of aromatic substitution patterns

emitted wavelengths can be observed in the scattered radiation (see Fig. 4.49). This is caused by the possibility that the vibration levels in ground and excited level differ before and after excitation. These fine differences represent the same energy differences as compared to IR spectroscopy. Two types of shifts can be observed, the Stokes and Anti-Stokes radiation. As illustrated in Fig. 4.49), measuring the scattering radiation by known exciting primary radiation allows to identify the differences of vibration energies according to IR spectroscopy.

The interpretation of Raman spectra is comparable to the approaches introduced for IR spectroscopy. Therefore, all implications and molecular information deduced from IR spectra can be obtained by Raman spectroscopy. However, intensity differs partly significantly between IR and Raman absorptions and, consequently, the spectra are complementary but not identical as illustrated schematically in Fig. 4.50.

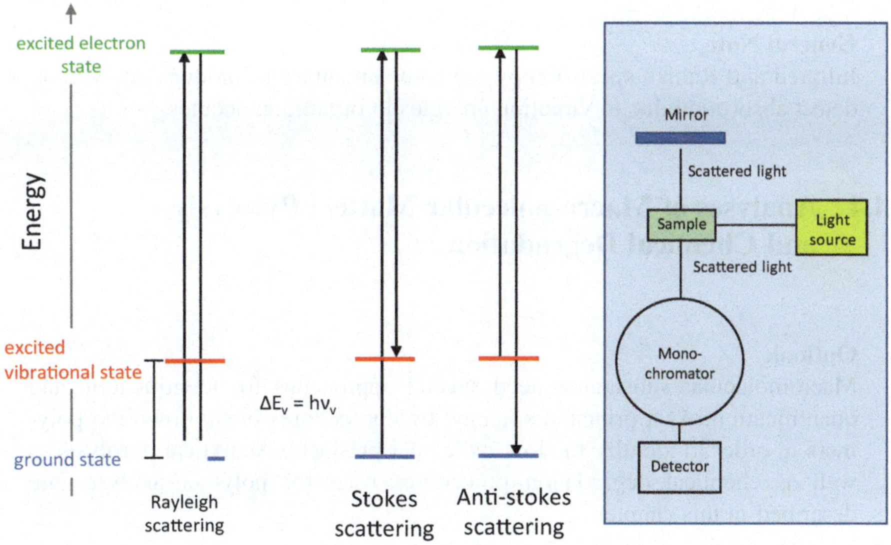

Fig. 4.49 Energy absorption scheme in Raman spectroscopy and the principal set-up of Raman spectrometer

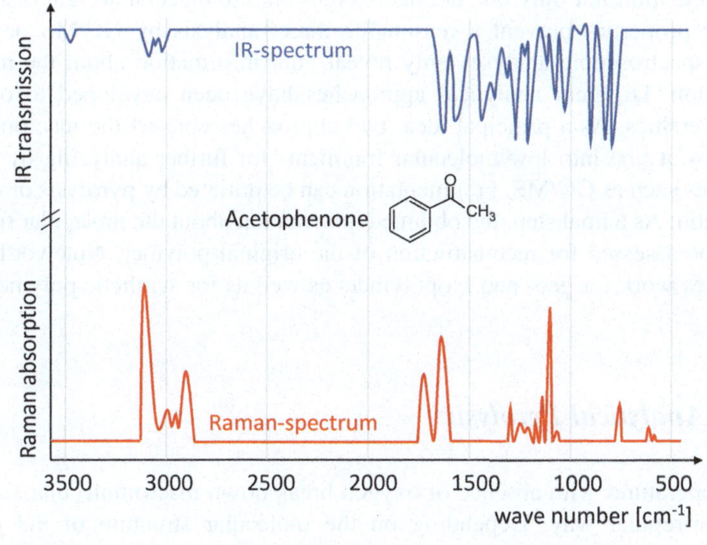

Fig. 4.50 Scheme of complementary IR and Raman spectra of acetophenone

General Note
Infrared and Raman spectroscopy are complementary techniques allowing to detect absorption due to vibration energies in organic molecules.

4.4 Analyses of Macromolecular Matter: Pyrolysis and Chemical Degradation

Outlook
Macromolecular substances need special approaches for identification and quantification. As a principal strategy, two procedures break down the polymers in order to identify the low molecular products. Analytical pyrolysis as well as chemical degradation procedures used for polymer analyses are described in this chapter.

The analysis of macromolecular substances needs some special considerations. Polymers are not volatile, consist of molecular repetition units in regular or irregular order and exhibit not only one defined weight but a molecular weight distribution. All these properties prevent a reasonable direct analysis by GC/MS or LC/MS, whereas spectroscopic analyses only reveal bulk information about the molecular composition. Different analytical approaches have been developed to overcome these adversities. As a principal idea, two approaches convert the macromolecular substances at first into low molecular fragments for further analyzing by standard procedures such as GC/MS. Fragmentation can be initiated by pyrolysis or chemical degradation. As a final step, the obtained information about the molecular fragments need to be assessed for reconstruction of the original polymer. Noteworthy, these approaches work for geo- and biopolymers as well as for synthetic polymers.

4.4.1 Analytical Pyrolysis

High temperatures with absence of oxygen break down macromolecular substances in a semi-regular way. Depending on the molecular structure of the polymer, pyrolysis produces a wide range of products from a few up to hundreds of fragments. Noteworthy, the overall pyrolysis yield of detectable compounds is commonly in the lower percentage level. As a principle approach, structure specific pyrolysis products need to be identified for both identification and quantification.

This can be illustrated best for synthetic polymers. In Fig. 4.51 indicative pyrolysis products are illustrated pointing more or less accurate to the original polymer structures of polyacrylamides, polyethylene or polystyrene. Similar

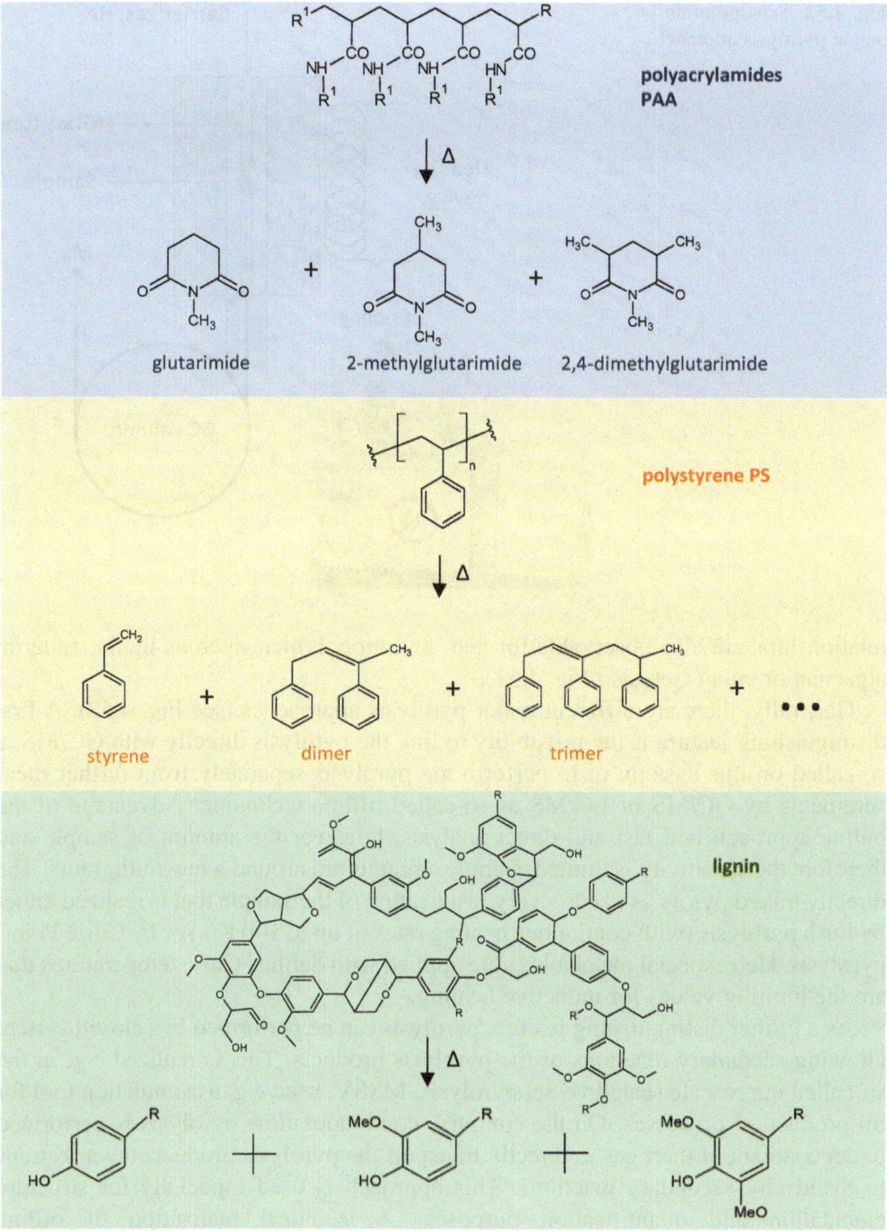

Fig. 4.51 Polymers and their specific pyrolysis products

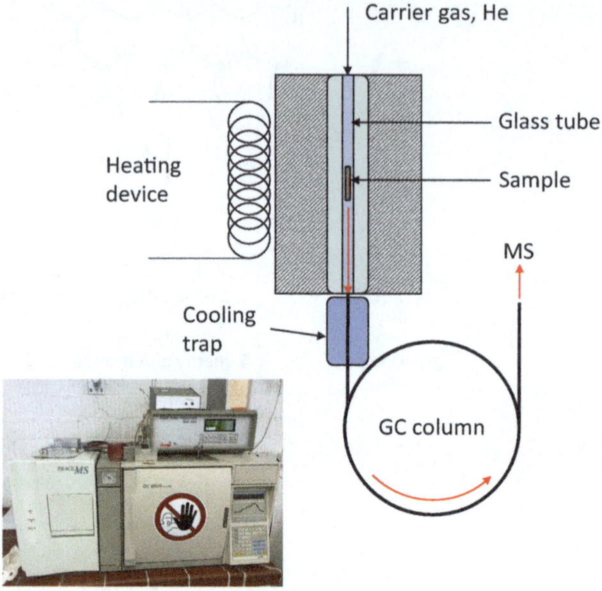

Fig. 4.52 Scheme of an online pyrolysis approach

relationships are also observable for geo- and biopolymers such as lignin, suberin, algaenan or cutan (see also Fig. 4.51).

Generally, there are different major pyrolysis approaches (see Fig. 4.52). A first distinguishing feature is the possibility to link the pyrolysis directly with GC/MS, a so-called on-line system, or to perform the pyrolysis separately from further measurements by GC/MS or LC/MS as so-called offline technique. Advantage of the online approach is a fast and direct analysis, however the amount of sample and therefore the sensitivity is limited, sample amounts are around a few milligrams. The directly linked pyrolysis needs a very fast heating of the sample that is realized either by flush pyrolysis (with continuous heating rates of up to 100 K/s) or by Curie Point-pyrolysis. Here, special metal alloys are applied with defined Curie temperatures that are the limiting values for inductive heating.

As a further distinguishing feature, pyrolysis can be performed in a closed system allowing secondary reactions of the pyrolysis products. This is realized e.g. in the so-called microscale sealed vessel pyrolysis, MSSV, used e.g. as simulation tool for oil production processes. On the contrary, continuous-flow pyrolysis is performed under a stream of inert gas to directly transport the pyrolysis products towards traps to avoid any secondary reactions. This approach is used especially for structure elucidation and quantification purposes. A technical realization of offline continuous-flow pyrolysis is visualized in Fig. 4.53.

For all pyrolysis systems the main parameters influencing the yield as well as the product spectra are the pyrolysis temperature and the pyrolysis time. Typically, temperatures between 450 and 900 °C are used mainly depending on the thermal maturity and thermodynamic stability of the macromolecules. The pyrolysis time varies for online and offline approaches. In online pyrolysis commonly a few

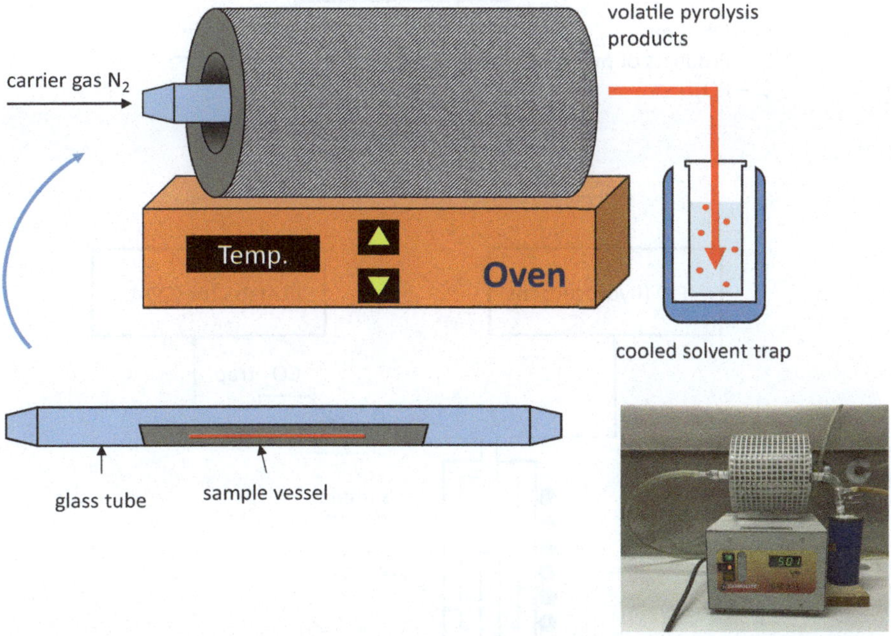

Fig. 4.53 Scheme of an off-line continuous-flow pyrolysis system

seconds are applied whereas in offline systems pyrolysis lasts between a few up to 30 min.

A special form of organic-geochemical analyses applies pyrolysis in the presence of water, so-called hydropyrolysis. Bearing in mind that in natural geological conditions water is almost always present, hydropyrolysis reflects better natural geological conditions and is, therefore, often used as simulation tool for geological processes, in particular catagenesis.

General Note
Independent from the technical approach (offline/online, sealed or continuous-flow) analytical pyrolysis produces specific products that can be analyzed for identification and quantification of polymers by GC/MS or LC/MS.

4.4.2 Bulk Pyrolysis

Beside its usage as molecular structure elucidating tool, pyrolysis can also be used for the characterization and quantification of the bulk organic matter. In Organic Geochemistry two common approaches are based on pyrolysis, the TOC analyses as well as Rock-Eval-pyrolysis.

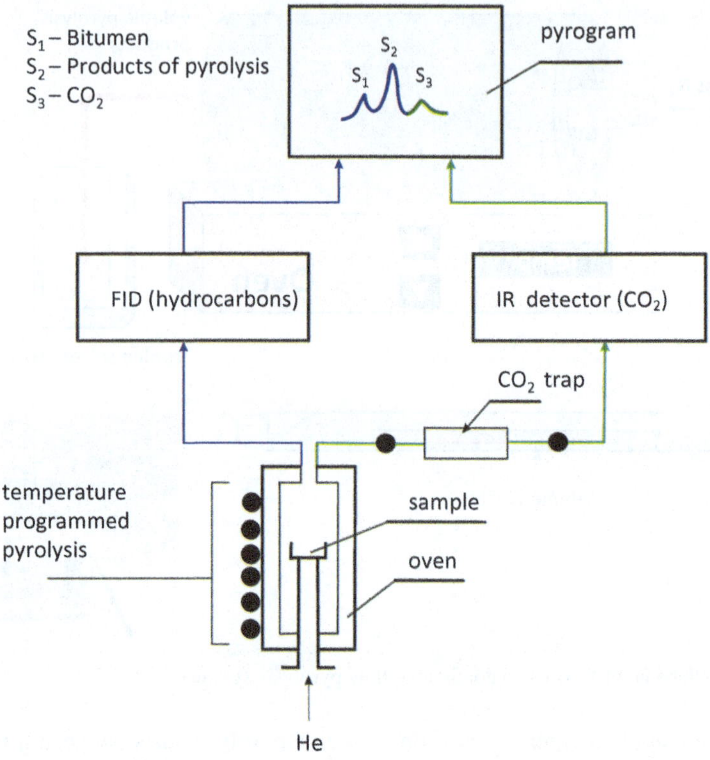

Fig. 4.54 Scheme of a Rock-Eval pyrolyzer

The determination of total organic carbon (TOC) is a fundamental parameter for a fast and rough characterization of samples. In contrast to analytical pyrolysis, that is exclusively performed under anaerobic conditions, TOC measurement is a combustion process heating up the sample material in presence of oxygen. This combustion converts organic matter more or less completely to CO_2, H_2O and further inorganic gases (H_2S, NO_2 ….), however, the organically bound carbon is converted to CO_2 that is measured easily by a simple IR detector (see Sect. 4.3.3). The differentiation of organic and inorganic carbon (e.g. in carbonates) is related to the combustion temperature. Since inorganic carbon is mainly more stable during combustion than organic one, first the TOC can be measured at temperatures up to approx. 550 °C, thereafter inorganic carbon will be released by temperatures up to 1000 °C. The total amount of TOC is finally calculated by the initially amount of sample and the measured CO_2 amount as analyzed by the calibrated IR detector.

A more precious information about the bulk organic matter can be obtained by the so-called Rock-Eval pyrolysis (see Fig. 4.54) introduced in 1977 by the French geochemist J. Espitalié. Here, the organic material is pyrolysed under special conditions. In a pyrolyzer, the untreated sample with known amount and TOC value is heated at a constant rate in a stream of inert gas (normally helium) within

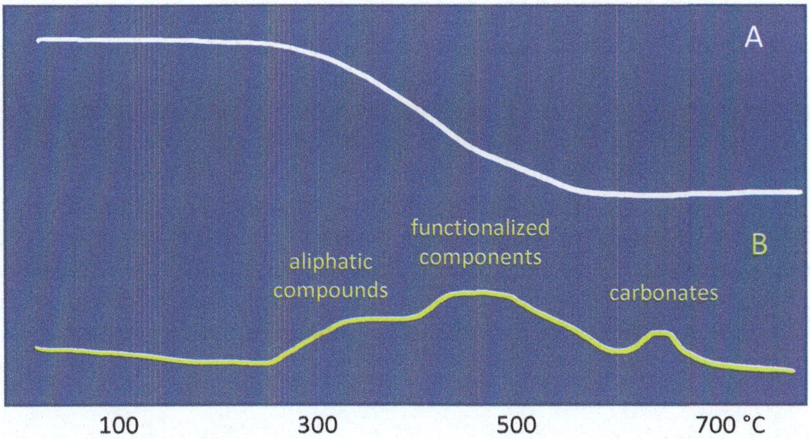

Fig. 4.55 Resulting curves from thermogravimetry DT (**a**) and differential thermogravimetry DGT analyses (**b**) of an oil shale sample

a certain temperature range, up to around 600 °C. The device is technically designed to detect three peaks during the heating period which are denoted by S1, S2 and S3. The peaks S1 and S2 are detected by an FID (flame ionization detector, see Sect. 4.1.1), that allows to trace released hydrocarbons and further organic substances. The last peak (S3) is an IR detector (similar to the TOC approach) that analyzes CO_2. This CO_2 is trapped during the pyrolysis process and deliberated after the heating period for further measurement.

The size of the first peak is proportional to the amount of native bitumen or thermodesorbable compounds. Peak S2 is proportional to the quantity of liquid products, which are produced in the pyrolysis of kerogen, representing the macromolecular part of the total organic matter. Finally, S3 represents amount of CO_2 released in the pyrolysis process and characterizes the organically bound oxygen as an indicator for functional groups in the organic material.

The resulting S1, S2 and S3 values are used to calculate specific parameters such as HI (hydrogen index) and OI (oxygen index) indices, that allow a classification of the kerogen type. Further on, the temperature at the maximum in peak S2 is named T_{max} and is used as indicator for the thermal maturity of the material.

Finally, there is a third pyrolysis-based approach to characterize the bulk organic matter, the thermogravimetry TG and its mathematical variation, differential thermogravimetry (DTG) analyses. Since this approach can be applied to a high variety of sample types, the technique is used in different disciplines, beside Organic geochemistry particularly material sciences.

A defined quantity of sample (typically up to 10 mg) is heated in the temperature range 30–900 °C, at a rate of about 5 °C/min under a nitrogen stream, and a flow rate of about 30–40 cm/min. During the pyrolysis, the weight of the sample is continuously measured. The result of the analyses are weight curves that have different shapes. Examples are given in Fig. 4.55.

Descending of a TG analysis curve is proportional to the quantity of products released during the thermal degradation of the sample. The weight loss is directly related to the mass loss during the pyrolysis of the organic matter. DTG analysis is somewhat more accurate. In this analysis, the height of peaks indicates the weight loss. However, DTG curve shows more maxima allowing a more specific interpretation. The total broad signal corresponds to the yield of the pyrolysis of organic matter. However, a fine evaluation allows to discriminate hydrocarbons from functionalized components characterizing the overall chemical composition of the organic material (see Fig. 4.55). Smaller peak corresponds to the degradation of other components in the sedimentary rock. For example, during analyses of an oil shale it probably refers to the degradation of carbonates.

4.4.3 Chemical Degradation

Break down of macromolecular structures for further analysis can be also achieved by chemical methods. The general approach is similar to pyrolysis, the degradation products can be used for structure characterization but only very sporadically for quantification purposes.

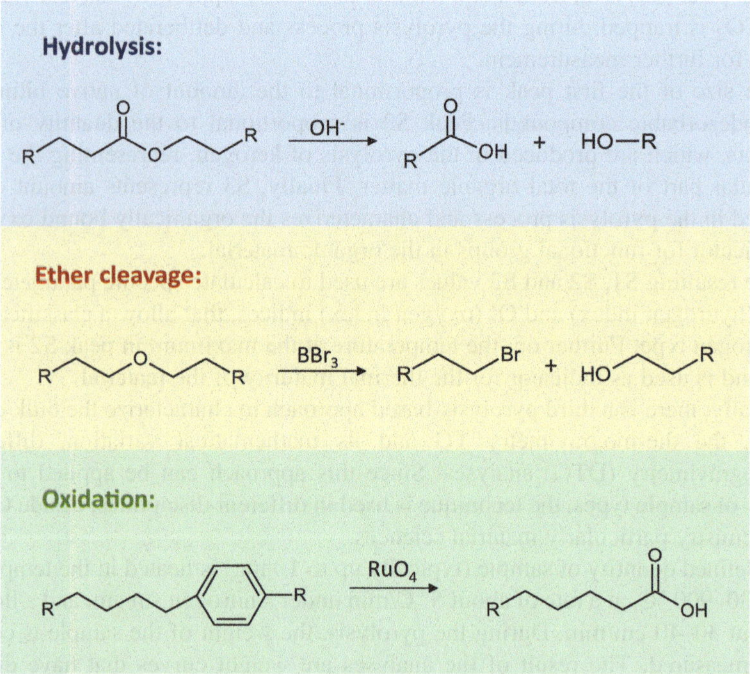

Fig. 4.56 Principal reactions of chemical degradation procedures

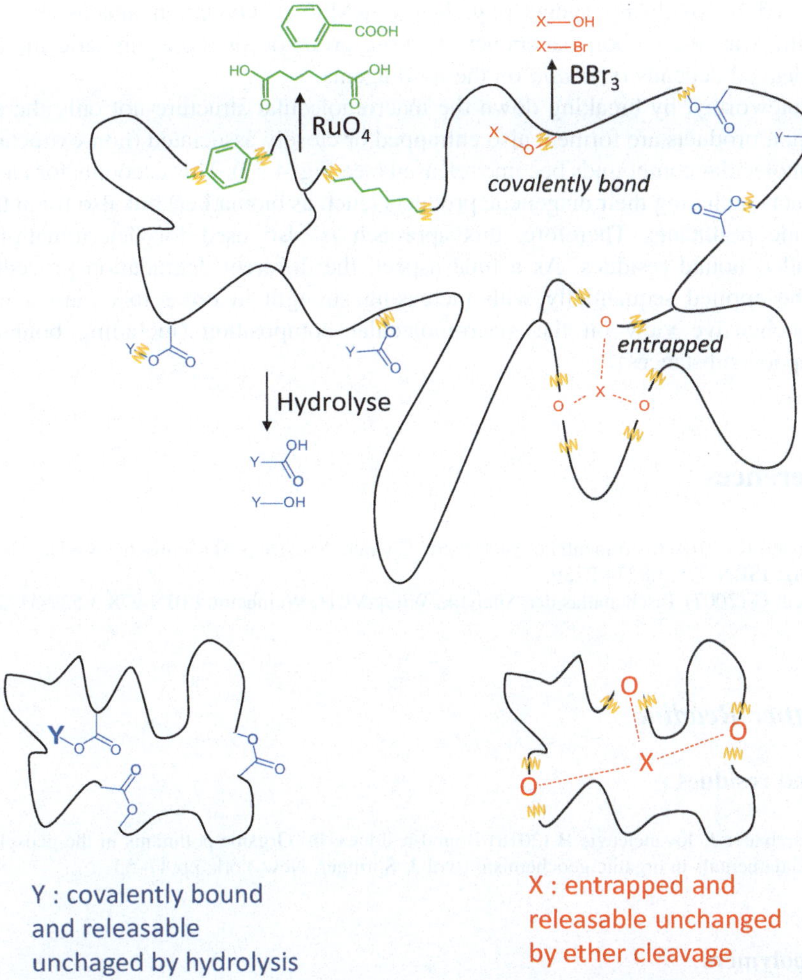

Fig. 4.57 General scheme on how degradation products and entrapped substances are released by chemical degradation

Using selective chemical degradation methods allows to distinguish the different bindings within the macromolecular structure but also the strength of linkage (see Fig. 4.56). The most common chemical degradation method is a soft one, the hydrolysis. Both acidic and alkaline hydrolysis are used to separate dominantly ester, amide or acetal bonds and can be applied to both biomacromolecules (peptides, polysaccharides …) and geopolymers (lignite, humic substances, kerogen ….). Degradation products cover alcohols, amides and carboxylic acids. A medium intensive chemical degradation can be performed with Lewis acids such as boron tribromide BBr_3 or aluminum trichloride $AlCl_3$. Hereby, ether linkages are cleaved forming alcohols but also bromides or chlorides. Finally, strong degradation can be achieved by oxidizing agents (e.g. RuO_4, $KMnO_4$). Oxidation attacks in a less specific way the carbon backbones. If more aromatic or more aliphatic moieties are cleaved depends *inter alia* on the used agents.

Noteworthy, by breaking down the macromolecular structure not only the degradation products are formed, also entrapped or closely associated (non-extractable) low molecular compounds become released (see Fig. 4.57). This accounts for natural products including their diagenetic products (such as biomarker) but also for anthropogenic pollutants. Therefore, this approach is also used for determination of so-called bound residues. As a final aspect, the different degradation procedures can be applied sequentially with increasing strength in order to obtain a more comprehensive view on the macromolecular composition (including bound or entrapped substances).

References

Cammann K (2010) Instrumentelle Analytische Chemie. Spektrum Akademischer Verlag, Heidelberg. ISBN 978-3-8274-2739-7

Schwedt G (2007) Taschenatlas der Analytik. Wiley-VCH, Weinheim. ISBN 978-3-527-31729-5

Further Reading

Bound residues

Schwarzbauer J, Jovancicevic B (2018) Bound residues. In: Organic pollutants in the geosphere. Fundamentals in organic geochemistry, vol 3. Springer, New York, pp 47–53

Geopolymers

Schwarzbauer J, Jovancicevic B (2016) Macromolecules. In: From biomolecules to chemofossils. Fundamentals in organic geochemistry, vol 2. Springer, New York, pp 127–158

Biomarker

Schwarzbauer J, Jovancicevic B (2016) The biomarker approach. In: From biomolecules to chemofossils. Fundamentals in organic geochemistry, vol 2. Springer, New York, pp 15–25

Chapter 5
GC/MS Data Evaluation

5.1 Identification

Outlook
The hyphenated system of gas chromatography-mass spectrometry GC/MS is a key technique for identification of organic substances in complex mixtures and at low concentration levels. Here, the key aspects of mass spectra data interpretation and the correlation with gas chromatographic information are presented for structure elucidation of organic analytes. Further on, some examples from environmental as well as fossil matter analyses are given.

Fragmentation is the most important process that allows to obtain information on the molecular structure of the analyte measured by mass spectrometry. Additionally, the separation by gas chromatography also provides information and contributes to an unambiguous identification of organic substances. The process of identification is nowadays supported by automated search algorithms, however a profound identification needs an in-depth knowledge of interpretation of both mass spectrometric and gas chromatographic data. Therefore, in a first subchapter the principal aspects of MS data interpretation are given followed by a brief description of up-to-date processes of identification and, finally, by some examples from environmental as well as fossil matter analyses.

J. Schwarzbauer, B. Jovančićević, *Introduction to Analytical Methods in Organic Geochemistry*, Fundamentals in Organic Geochemistry,
https://doi.org/10.1007/978-3-030-38592-7_5

5.1.1 *Short Course of Mass Spectra Interpretation and GC/MS Based Identification*

Key aspect of mass spectra interpretation is related to the fragmentation processes occurring in the ion sources but that are mainly forced by the molecular properties of the analytes. Interpretation of mass spectra uses various approaches.

A first one is the interpretation of the molecular ion. As a very simple fact, the mass of organic molecules is commonly an even value (with one exception, described later as the *nitrogen rule*). Consequently, fragments derived from one bond linkage exhibit even mass values. The resulting mass spectra exhibit an even molecular ion and are dominated by even fragment signals. Therefore, the molecular ion can be characterized as highest even m/z value. The intensity of the molecular characterizes the stability of the molecule or, with other words, the potential to resist fragmentation from electron bombardment in the ion source. As a simple rule, the higher the molecular peak intensity the higher the thermodynamic stability of the molecule. Noteworthy, for very unstable molecules the molecular peak can disappear and is not visible in the mass spectrum. In that cases, only even fragment ions occur in the spectra. Examples are given in Fig. 5.1.

A second approach tries to distinguish specific patterns. This can be well illustrated by examine mass spectra of aliphatic and aromatic hydrocarbons.

In Fig. 5.2a the mass spectra of *n*-tetradecane is given. A distinct pattern is obvious that is typical for *n*-alkanes in general. A higher number of fragments, all with the same m/z differences, appear with an abundance maximum at lower masses (m/z 43 and 57 corresponding to C_3H_7 and C_4H_9 ions) followed by an exponential but uniform decrease of abundance to higher masses. Only a slight increase towards the molecular ion is visible. The relative low abundance of the molecular ion points to a low thermodynamic stability of molecule or a high intrinsic potential for fragmentation. The pattern of peak intensities also provides information on the stability of the fragmented moieties: the higher the abundance the higher the thermodynamic stability. Hence, propyl and butyl ions seem to be the most stable ones. The systematic differences between the main peaks of m/z 14 clearly reflect the homologues units of methylene groups $-CH_2-$. Interestingly, the insertion of a double bond does not change the principal pattern, but shifts the homologues mass series by two units, reflecting the two missing hydrogen atoms (see Fig. 5.2b). Noteworthy, the position of the double bond cannot be determined by mass spectrometry due to a dislocation of the unsaturation along the carbon chain during ionization.

A first glance on the quality of MS based identification can be obtained by comparing the mass spectra of the *n*-alkane (Fig. 5.2) with those of branched alkanes, so-called *iso*-alkanes as given in Fig. 5.3. Variety of fragment peaks as well as the systematic mass differences are obviously identical, but the abundance pattern varies slightly. A few peaks in the higher mass range exceed a little bit the uniform exponential decrease. These shifts point to the branched moieties, in particular to their positions within the carbon chain. As basic aspect, the bond

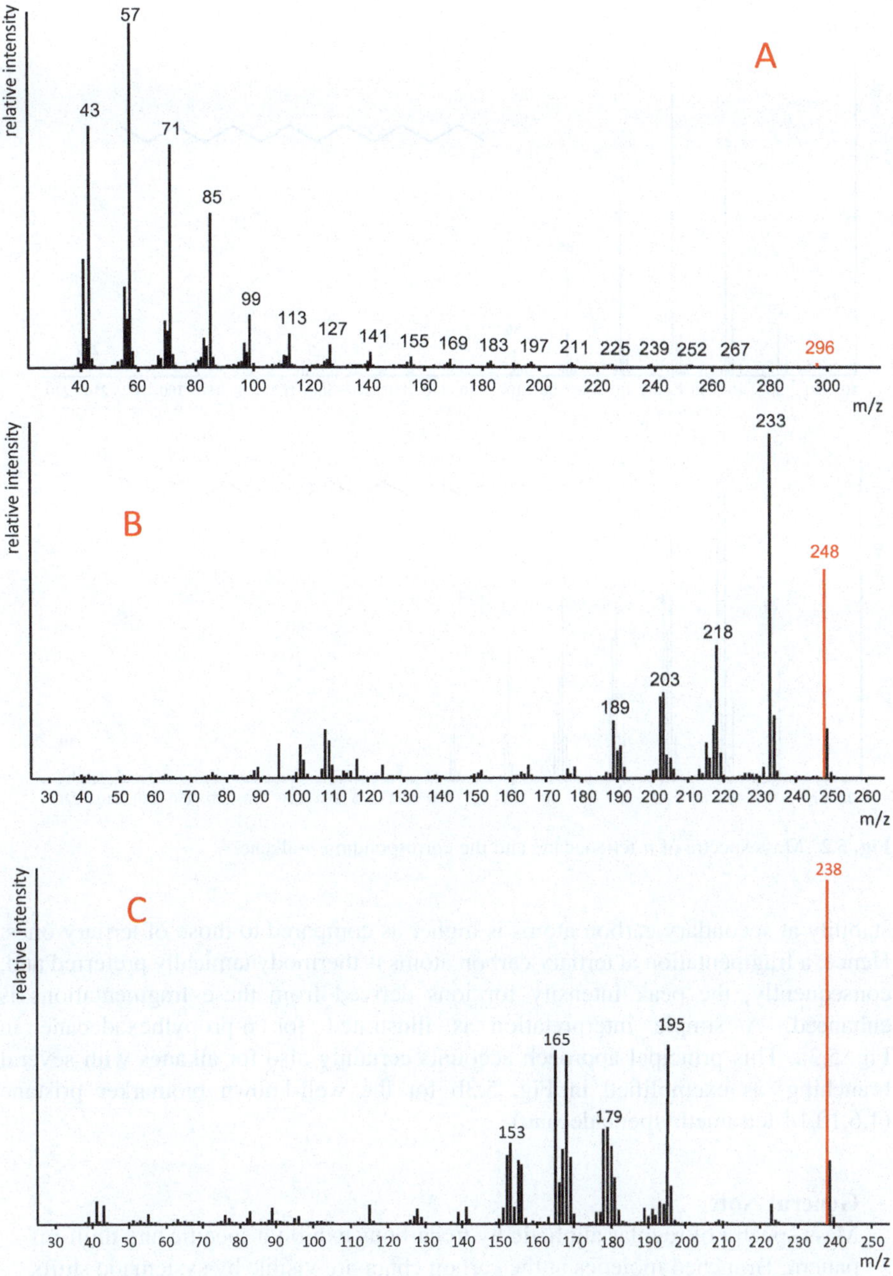

Fig. 5.1 Variation of molecular peak intensities as the result of molecule stability. (**a**) 8-Isopropyl-1,3-dimethylphenanthrene; (**b**) heneicosane; (**c**) 9-butyl-1,2,3,4-tetrahydroanthracene

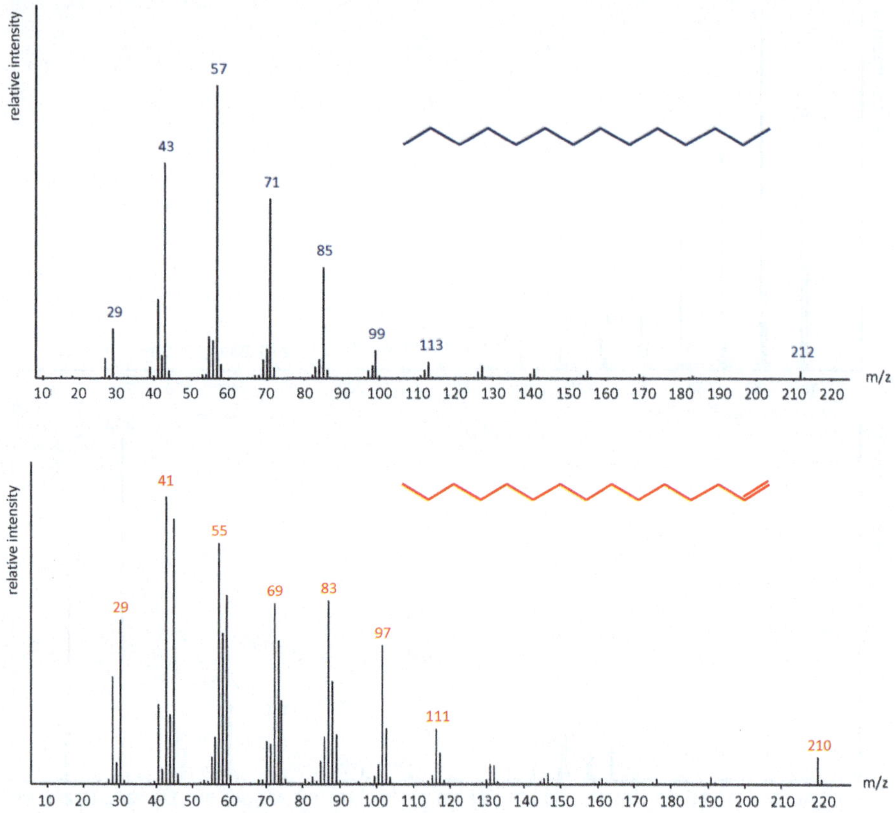

Fig. 5.2 Mass spectra of *n*-tetradecane and the corresponding *n*-alkene

stability at secondary carbon atoms is higher as compared to those of tertiary ones. Hence, a fragmentation at tertiary carbon atoms is thermodynamically preferred and, consequently, the peak intensity for ions derived from these fragmentations is enhanced. A simple interpretation is illustrated for 6-propylhexadecane in Fig. 5.3a. This principal approach accounts certainly also for alkanes with several branching, as exemplified in Fig. 5.3b for the well-known biomarker pristane (4,6,10,14-tetramethylpentadecane).

General Note

Mass spectra of regular aliphatic hydrocarbons exhibit a specific and uniform pattern. Branched moieties at the carbon chain are visible by systematic shifts in the relative abundances.

Following the approach of mass spectra interpretation by patterns, a distinct difference can be observed between mass spectrometric properties of aliphatic and

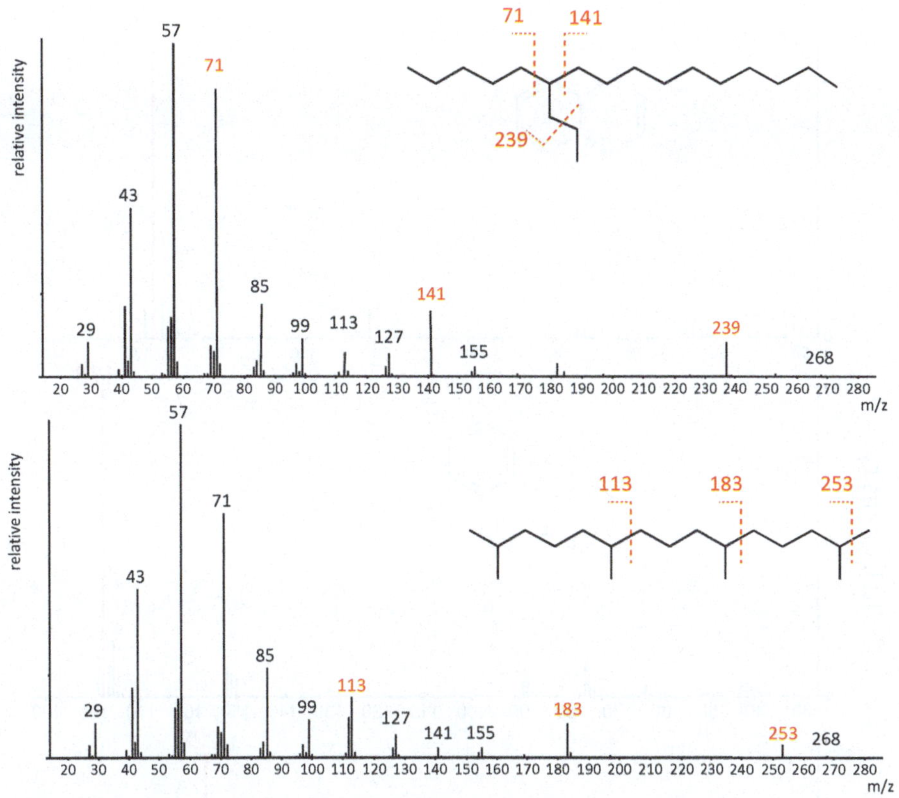

Fig. 5.3 Mass spectra of 6-propylhexadecane and pristane with specific fragmentations

aromatic hydrocarbons. As mentioned, aliphatic mass spectra are characterized by numerous and systematic occurring peaks with higher intensities at lower masses. On the contrary, mass spectra of aromatic hydrocarbons exhibit only very scarce peaks but with high intensities dominantly for the molecular ion as illustrated for some aromatics in Fig. 5.4.

However, often not solely aliphatic or aromatic moieties exist in organic molecules but both types of hydrocarbon moieties appear together. A close look on the corresponding mass spectra allows to distinguish the individual structural units. Interestingly, at the connection of the aromatic and aliphatic units not the direct bond is the preferred point of fragmentation but the proximate aliphatic bond. The formed methylene-substituted aromatic ions are stable as the result of enhanced mesomerism. For phenyl substitution the corresponding benzyl fragment at m/z 91 is characteristic. In Fig. 5.5 this preferred fragmentation is obvious by comparing the base peaks of a methylated and an ethylated aromatic hydrocarbon. For the first one, the aromatic pattern remains, whereas in the latter case a preferred separation of the methyl group is obvious that produces a base peak at M-15 m/z.

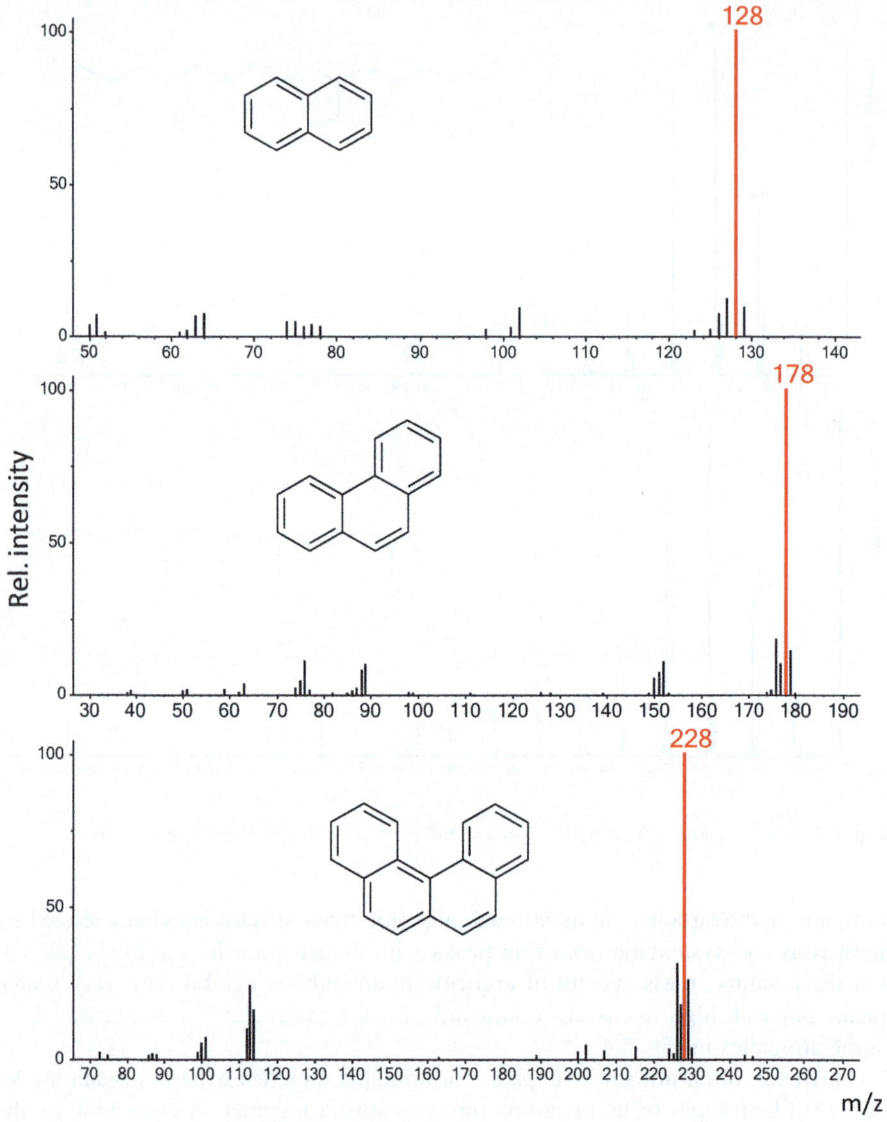

Fig. 5.4 Mass spectra of naphthalene, phenanthrene and benz(c)anthracene

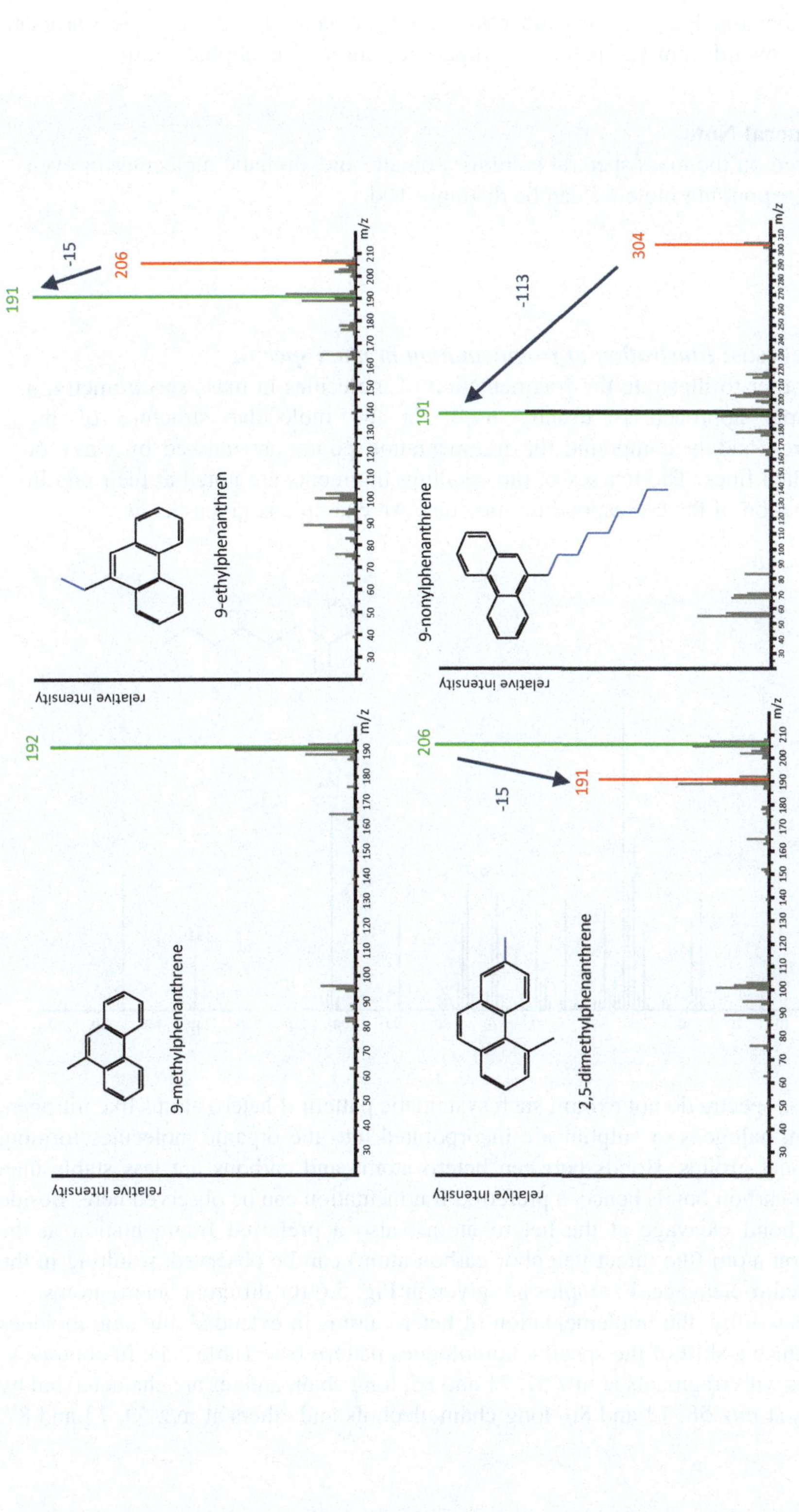

Fig. 5.5 Mass spectra of unsubstituted phenanthrene and some methyl-, ethyl- and nonylsubstituted derivatives with specific fragmentations reflecting the superimposition of aliphatic and aromatic pattern

Further on, Fig. 5.5 also illustrates the systematic shift from a pure aromatic pattern towards a mixed pattern by superimposition of an aliphatic unit.

General Note
Based on the mass spectral patterns aromatic and aliphatic molecules or even corresponding moieties can be distinguished.

Excursus: ***Illustration of fragmentation in mass spectra***
In order to illustrate the fragmentation of molecules in mass spectrometry, a simple approach is usually used. In the molecular structure of the corresponding compound the fragmentation points are marked by wavy or dashed lines. The masses of the resulting fragments are noted at their end in direction of the corresponding moieties. An example is given below:

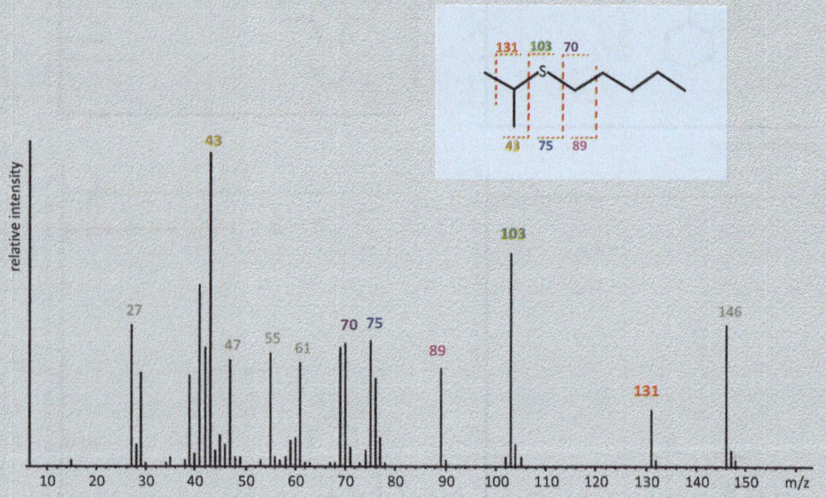

Mass spectra do not exhibit such systematic pattern if hetero atoms like nitrogen, oxygen, halogens or sulphur are incorporated into the organic molecules forming functional groups. Bonds between hetero atoms and carbons are less stable than carbon-carbon bonds hence, a preferred fragmentation can be observed here. Beside direct bond cleavage at the hetero atoms, also a preferred fragmentation at the α-carbon atom (the direct neighbor carbon atom) can be observed, resulting in the so-called α-cleavage. Examples are given in Fig. 5.6 for different hetero atoms.

Noteworthy, the implementation of hetero atoms in extended aliphatic moieties can induce a shift of the specific homologues pattern (see Table 5.1). In contrast to alkanes with fragments at m/z 57, 71 and 85, long chain amines are characterized by signals at m/z 58, 72 and 86, long chain alcohols and ethers at m/z 59, 73 and 87.

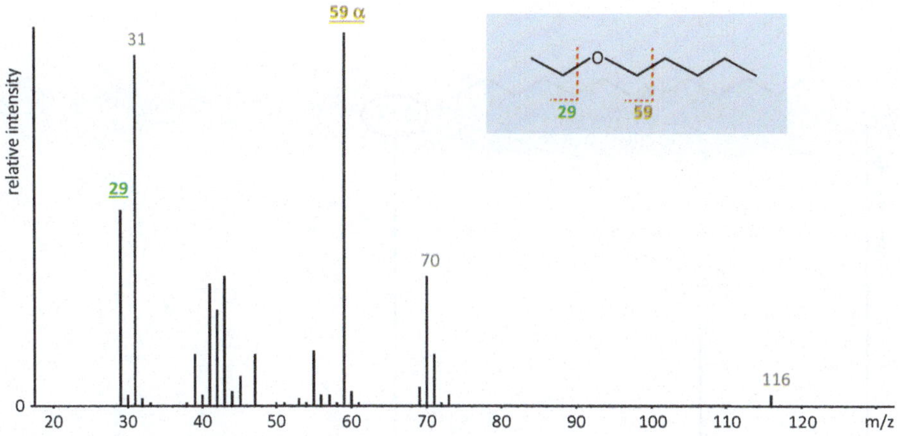

Fig. 5.6 Mass spectra of ethylpentyl ether representing preferred fragmentation at hetero atoms and in α-position

Table 5.1 Significant ion series

Functional group	Start mass	Ion series
Alkanes	29 ($C_2H_5^+$)	43, 57, 71, 85, 99, 113, ...
Alkenes	27 ($C_2H_3^+$)	41, 55, 69, 83, 97, 111, ...
Ethers, alcohols	31($H_2C{=}O^+H$)	45, 59, 73, 87, 101, ...
Ketones	43 (H_2C-CO^+)	43, 57, 71, 85, 99, 113, ...
Amines	30 ($H_2C{=}N^+H_2$)	30, 44, 58, 72, 86, 100, ...

These changes can be highly indicative. However, it fails especially for some carbonyls such as ketones and aldehydes, since the C=O group exhibits the same mass as compared to C_2H_4, consequently no shift can be observed.

There is one more special aspect that sometimes also provides insight into the molecular composition, the so-called nitrogen rule. As a very simple fact, the mass of organic molecules is commonly an even value. Consequently, fragments derived from one bond linkage exhibit even mass values. The resulting mass spectra exhibit an even molecular ion and are dominated by even fragment signals. Only in case of fragment formation based on two bond linkages (e.g. eliminations) some even fragment ions appear. This principal design of mass spectra gets reversed if nitrogen appears in the molecule. To be accurate, if a molecule exhibits an odd number of nitrogen atoms, the corresponding molecular mass is odd, and fragments formed by mono linkages are even (see Fig. 5.7). The corresponding mass spectra are inverse to most other spectra with odd molecular ions and mostly even fragment ions. This effect is described by the 'nitrogen rule'. Noteworthy, this rule influences also the identification of the molecular ion. It is not necessarily the highest even m/z value but can change under the circumstances described to the highest odd m/z value (if visible!).

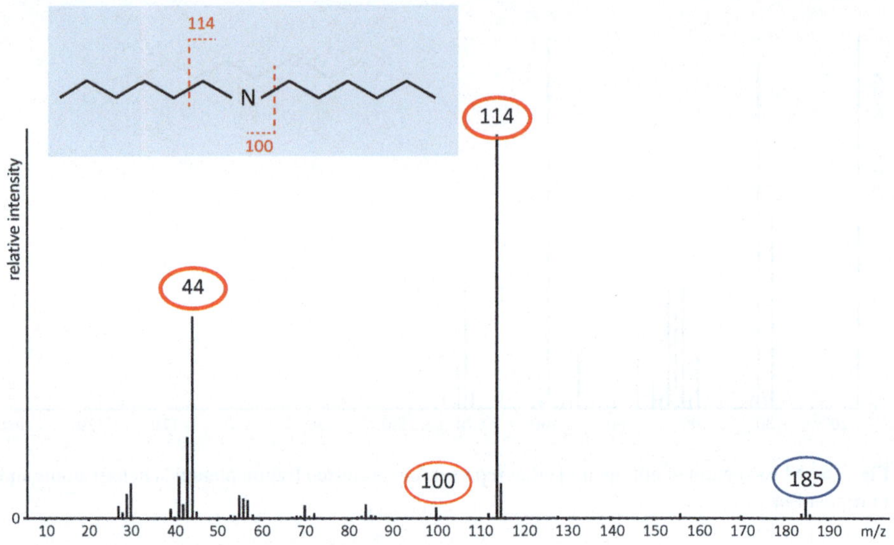

Fig. 5.7 Mass spectra of dihexylamine illustrating the nitrogen rule

All principal statements presented so far are relatively simple. Normally, in more complex mass spectra superimposition of different aspects are observable. This makes interpretation more complex but allows some deeper insights. This will be exemplified in the following by the interpretation of fatty acid methyl esters and its *iso*- and *anteiso*-isomers. In Fig. 5.8 indicative sections of mass spectra of unbranched pentadecanoic acid methyl esters and *iso*-methyl (methyl branched at the next-to-last position) and *anteiso*-methyl (methyl branched at the third-to-last position) substituted homologues are illustrated. All methyl esters have one common fragmentation by losing the Me-O moiety leading to fragments with M-31 m/z, here resulting in m/z 225. Accompanied there is the loss of ethyl units forming peaks with M-29 m/z, here as m/z 227. The relative abundance of these two peaks allow to distinguish the isomers, because the first fragmentation is not influenced by the isomerism. However, the fragmentation leading to m/z 227 is preferred in the *anteiso*-isomer due to branching at the corresponding position but nearly impossible for the *iso*-isomer. These circumstances induce an elevated 227/225 ratio for the *anteiso*-isomer and a strong depletion of this value for the *iso*-isomer as compared to the 227/225 ratio for the unbranched isomer.

A second example illustrating more complex mass spectrometric interpretation is related to a specific rearrangement, the so-called *McLafferty rearrangement*. It is also related to the mass spectrometric behavior of fatty acid methyl esters but focusses solely on the functional group (see Fig. 5.9). At the methyl ester group the aliphatic chain can be folded to a quasi-six ring and as a result a hydrogen atom shifts from the aliphatic side chain to the carbonyl oxygen. This unstable transition gets stabilized by elimination of an olefinic moiety and the formation of an unsaturated ester unit forming a very characteristic m/z 74 peak. This peak represents the key fragment for identification of methyl esters. A comparable rearrangement of

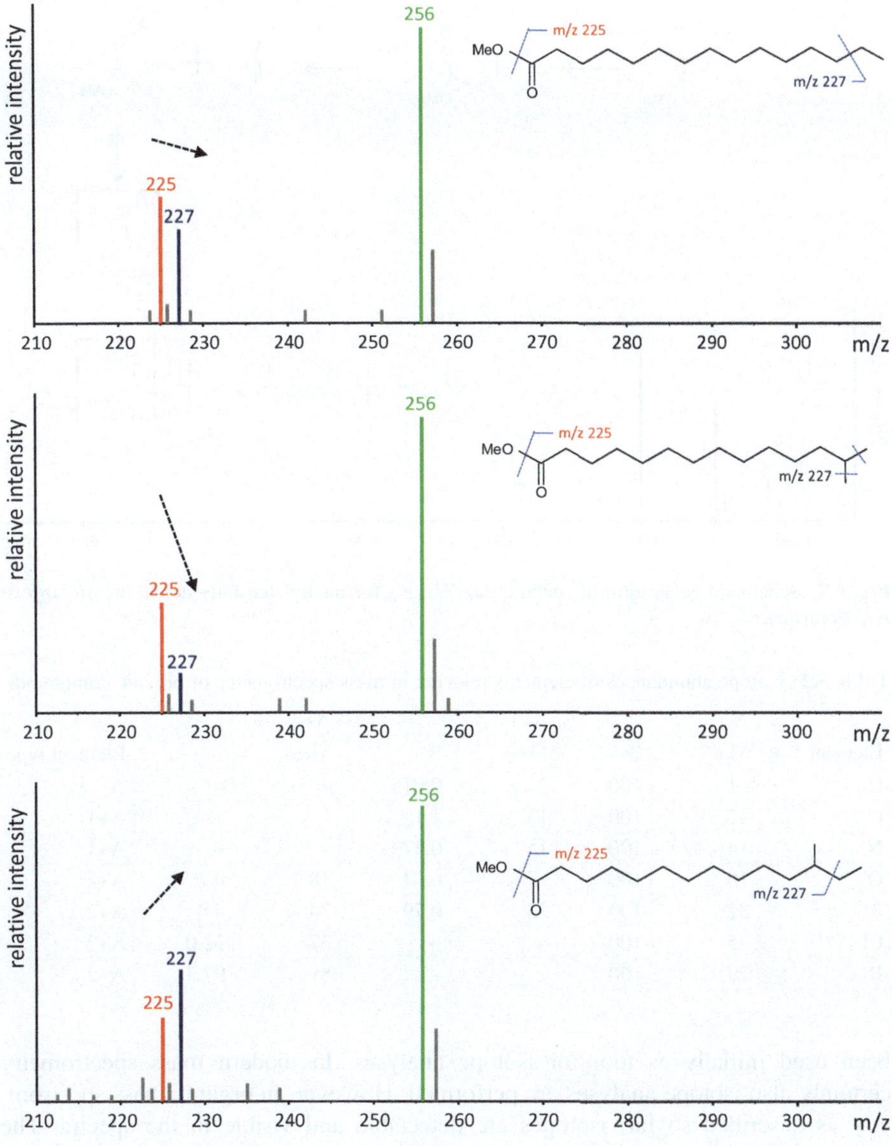

Fig. 5.8 Indicative fragmentation behavior of unbranched and methyl branched pentadecanoic acid methyl esters

methyl ketones forms also specific peaks, here at m/z 58. Noteworthy, from a formal point of view, this reaction is an elimination by cleaving two separate bonds, hence even fragments are produced that are easily to identify among all the commonly odd peaks resulting from fragmentations by solely one cleavage.

Finally, there is one main aspect that facilitates especially the identification of many organic pollutants. Going back to historical aspects, mass spectrometry has

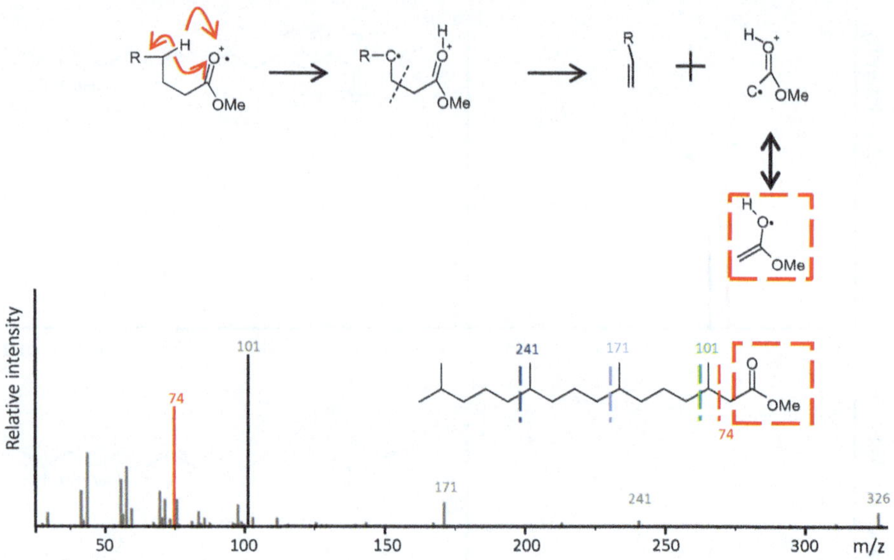

Fig. 5.9 Schematic generation of specific m/z 74 ions for methylated fatty acids, the *McLafferty rearrangement*

Table 5.2 Isotope abundances of elements relevant in mass spectrometry of organic compounds

Element	A		A+1		A+2		Element type
	Mass	%	Mass	%	Mass	%	
H	1	100	2	0.015	–	–	A
C	12	100	13	1.1	–	–	A+1
N	14	100	15	0.37	–	–	A+1
O	16	100	17	0.04	18	0.2	A+2
S	32	100	33	0.79	34	4.4	A+2
Cl	35	100	–	–	37	32.0	A+2
Br	79	100	–	–	81	97.3	A+2

been used initially as tool for isotope analyses. In modern mass spectrometry certainly also isotope analyses are performed. However, in organic mass spectrometry as described so far, isotopes are detectable and visible in the spectra. The corresponding elements cover mainly carbon and hydrogen as well as oxygen, sulphur, nitrogen and halogens. Noteworthy, for organic molecules only a few stable isotopes are of interest. To use the analytical potential of these isotopes, the relative abundance of the isotope signals needs to be high enough to unambiguously identify the isotopic signals. Beside the abundance, also the mass differences are significant as summarized in Table 5.2.

Most important element in organic chemistry is carbon, that exhibits one stable isotope in addition to ^{12}C, the ^{13}C isotope representing a so-called 'A+1' isotope. The relative abundance of ^{13}C is low with around 1%, hence this signal is not very

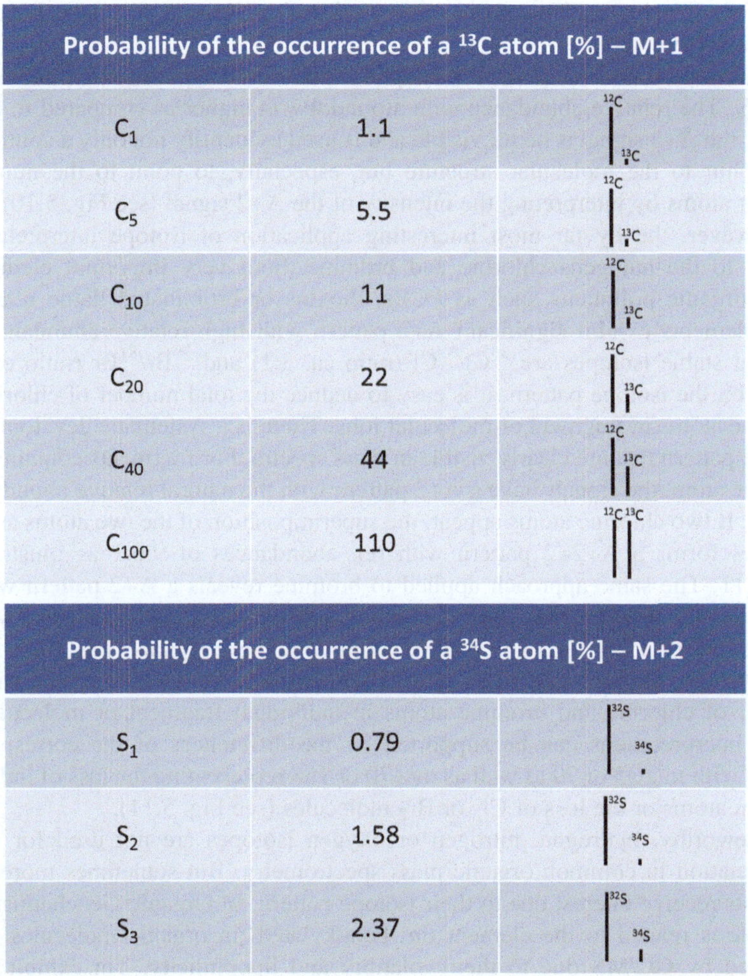

Fig. 5.10 Development of the relative intensities of the A+1 isotope signals of carbon (^{13}C) and A+2 signals of sulphur (^{34}S) by increasing atom numbers

significant. However, the low abundance of the signal will compensate by the number of carbon atoms in the organic molecule. This is illustrated in Fig. 5.10. The signal at A+1 (A represents any fragment or molecule ion) becomes of higher intensity in the area of around 10 carbon atoms in the molecule with intensities of approx. 10% and becomes dominant at around C_{100} with over 100% rel. abundance. This trend is related to the likelihood of occurrence of one ^{13}C atom in the molecule that increases with carbon atom number. Very few applications use the relative abundance of A and A+1 to quantify the number of carbon atoms in the analyte. However, this needs very accurate measurements and certainly the consideration of other A+1 isotopes e.g. from hydrogen (^{2}H, D), nitrogen (^{15}N) or sulphur (^{33}S).

Therefore, this approach remains as a very seldom application of isotope-based mass spectra interpretation.

A second interesting element is sulphur with a A+2 type isotope pattern of ^{32}S and ^{34}S. The relative abundance with around 4% is higher as compared to carbon. Hence, the A+2 signal is better visible and is used to identify not only a contribution of sulphur to the molecular structure but, especially, to point to the number of sulphur atoms by interpreting the intensity of the A+2 signal (see Fig. 5.10).

However, the by far most interesting application of isotope interpretation is related to the halogens chlorine and bromine, both very important elements in many organic pollutants such as PCBs, dioxins or brominated flame retardants. Both elements exhibit significant A+2 pattern with high relative abundances, the relevant stable isotopes are $^{35}Cl/^{37}Cl$ (ratio ca. 3:1) and $^{79}Br/^{81}Br$ (ratio ca. 1:1). Based on the isotope pattern, it is easy to deduce the total number of chlorine and bromine atoms in fragment or molecular ions. There is a systematic development of isotope pattern that are clearly visible in mass spectra. For fragments containing one chlorine atom, the signals have a A+2 pattern with the natural relative abundance of ca. 3:1. If two chlorine atoms appear, the superimposition of the two atoms and their isotopes forms a A+2+2 pattern with rel. abundances of 9:6:1 as illustrated in Fig. 5.11. The same approach applied to bromine reveals a A+2 pattern with 1:1 intensities for one bromine atom containing signals, but a A+2+2 pattern with rel. abundances of 1:2:1 for two bromine atom containing signals (Fig. 5.9). Consequently, it is obvious that these unique patterns can be used for calculating the number of chlorine and bromine atoms in individual fragment or molecule ions. These interpretations can be supported by the differences of the corresponding signals with m/z 35 or 79 as well as m/z 70 or 158 representing the loss of individual halogen atoms or the loss of Cl_2 or Br_2 molecules (see Fig. 5.11).

Noteworthy, hydrogen, nitrogen or oxygen isotopes are not used for spectra interpretation in common organic mass spectrometry. But sometimes more exotic elements receive interest due to their isotope pattern. In Organic Geochemistry one example is related to the element tin. Peralkylated tin organic molecules can be analyzed by GC/MS due to their volatility and lipophilicity. Tin exhibit a very unique pattern of stable isotopes with elevated relative abundances. Based on this pattern, tin organic compounds can be easily detected in mass spectrometry. This is illustrated in Fig. 5.12 for biomethylated tributyl tin, a formerly common ingredient in antifouling paints.

In summary, there are various aspects that can be used for an ab initio interpretation of mass spectra. These approaches cover the examination of the overall pattern for differentiating aliphatic and aromatic moieties, the influence of branching on fragment intensities, the interpretation of the molecular ion intensity, the consideration of preferred fragmentation at hetero atoms, the observation of even to odd fragment distribution and, finally, the identification of isotope pattern.

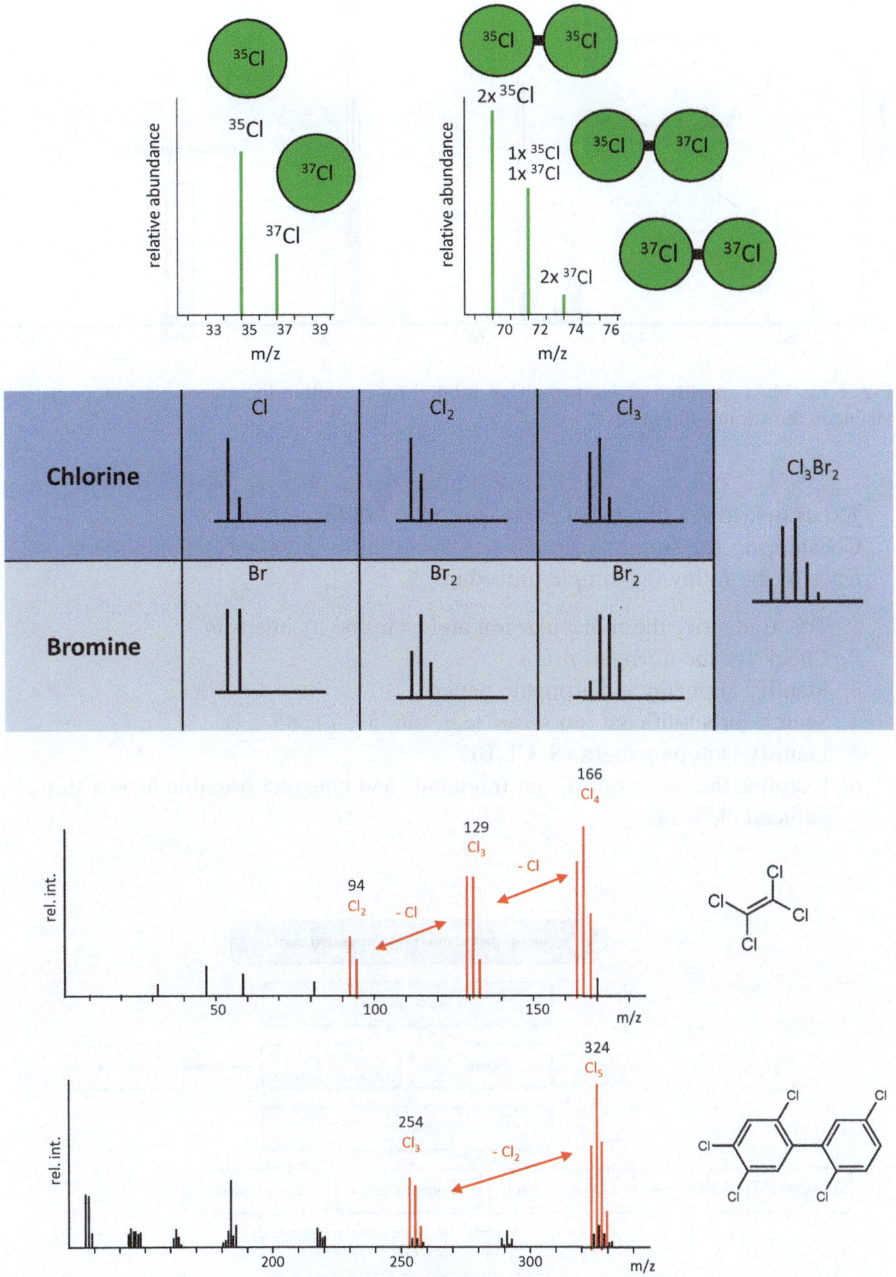

Fig. 5.11 Systematic isotope pattern of chlorine and bromine in mass spectrometry and their appearance in exemplary mass spectra of pollutants (tetrachloroethene and pentachlorobiphenyl)

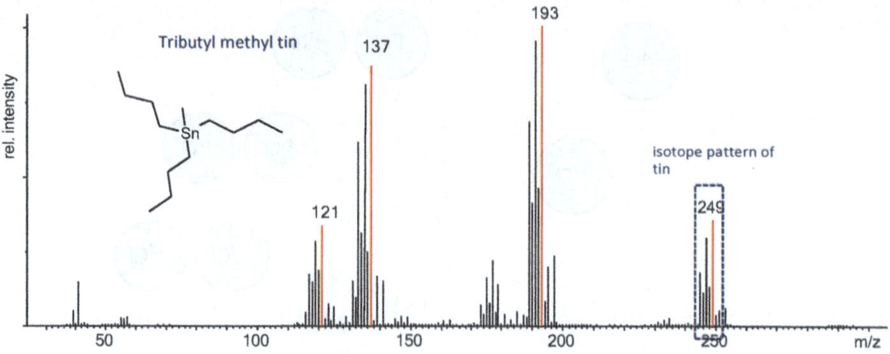

Fig. 5.12 Mass spectrum of biomethylated tributyl tin indicating the unique tin isotope pattern visible in the various fragments

Excursus: *Rules for initial mass spectrum interpretation*

Considering the summarized aspects of ab initio mass spectra interpretation leads to the following simple procedure:

1. Try to identify the molecular ion and examine its intensity
2. Check for the *nitrogen rule*
3. Identify aliphatic and aromatic pattern
4. Search for significant ion series (e.g. m/z 57, 71, 85 …)
5. Identify isotope pattern (S, Cl, Br)
6. Examine the more significant fragments and consider possible hetero atom induced cleavages

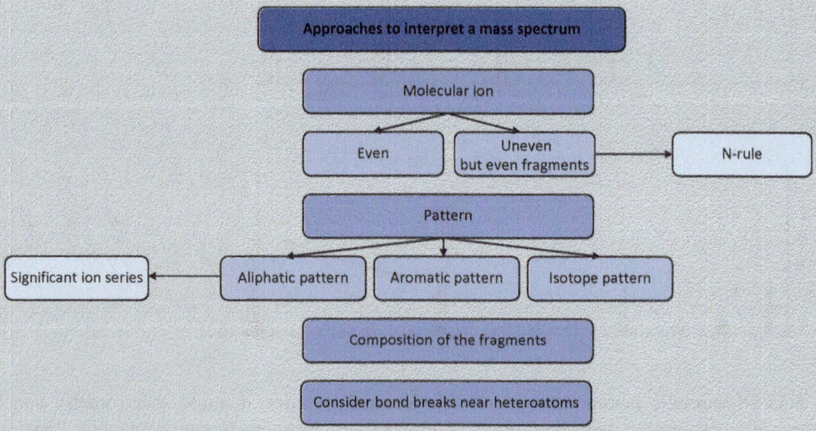

5.1.2 *Ion Chromatograms as a Key Feature for in Depth Examination*

A more special feature in GC/MS analyses allow a more detailed investigation of the data. Mass spectra are composed on the intensities of individual ions. The chromatogram obtained by GC/MS analyses consists of the total ion intensities (sum of all ions per mass spectrum) in relation to the retention time. The chromatogram consists of individual points, each of them representing one mass spectrum measured in a short time period (e.g. 0.5 s). The chromatogram is called total ion chromatogram (see Fig. 5.13) and at each point of the chromatogram the corresponding mass spectrum can be obtained. A variation of constructing a chromatogram is called ion chromatogram, considering the intensities of only one ion within all the measured mass spectra. This is also reflected in Fig. 5.13.

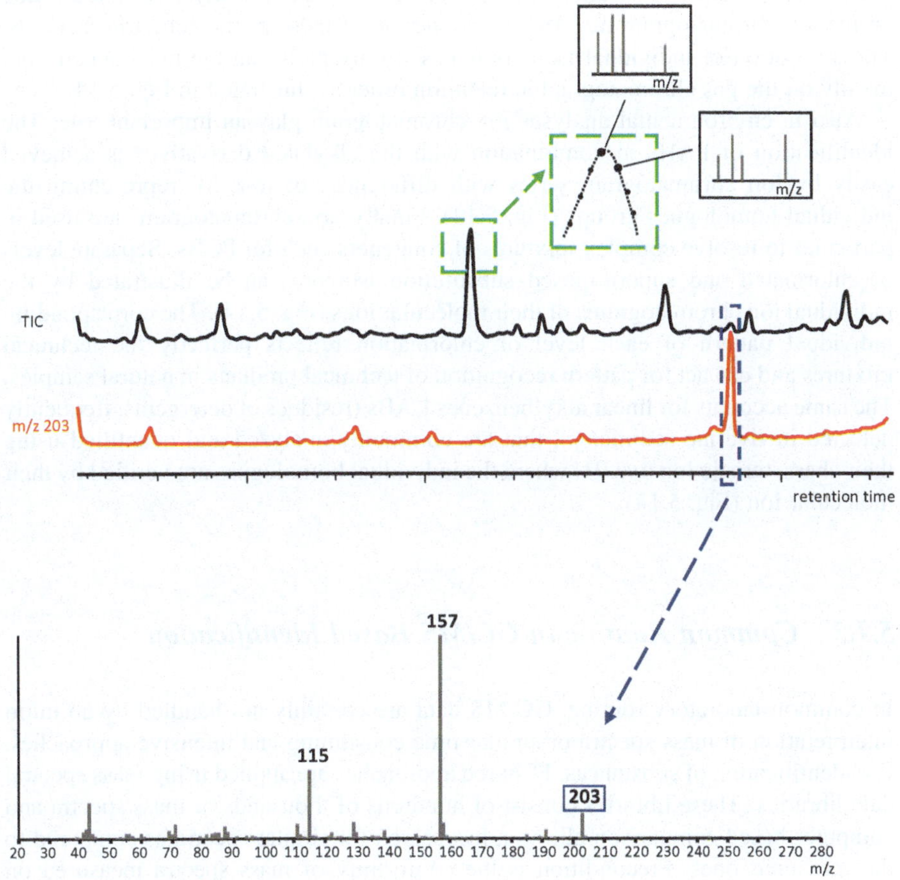

Fig. 5.13 Principals of ion chromatograms

The ion chromatogram allows some more special examinations. An important aspect is the quantification (see also Sect. 5.2). Ion chromatograms of specific ions (characterized e.g. by high masses, even values or unusual fragments) exhibit a better signal-to-noise ratio allowing to improve the sensitivity of the quantitative analyses (see Fig. 5.13 for m/z 203).

Ion chromatograms also allow a faster identification of preselected analytes. One just need to create one or more ion chromatograms derived from specific ions of the analytes' mass spectrum. At the position where all chromatograms exhibit a peak, most probably the wanted analyte can be identified. However, as a last but maybe most important feature, ion chromatograms allow to visualize isomers and homologue series. This is an important approach in Organic Geochemistry. As significant examples, analyses of fossil matter (coals, petroleum, kerogen etc.) comprise often *n*-alkane distributions as well as analyses of hopanes and steranes. The homologues series of *n*-alkanes can be easily detected by the ion fragment m/z 57 or 71 (or 81 . . .) and organic-chemically important parameter such as carbon preference indices (CPIs) can be directly visualized (see Fig. 5.14). Hopane analysis is often based on the ion chromatogram m/z 191 as unique ion of these pentacyclic triterpenoids. The attribution of individual isomers is based only partly on the mass spectra but mainly on the gas chromatographic retention order as illustrated in Fig. 5.14.

Also in environmental analyses ion chromatogram play an important role. The identification of PAHs in combination with the alkylated derivatives is achieved easily by ion chromatogram series with differences of m/z 14, representing the individual homologues groups (Fig. 5.14). Finally, ion chromatograms are used in particular to resolve complex mixtures of congeners such for PCBs. Separate levels of chlorinated and superimposed substitution isomers can be illustrated by the individual ion chromatograms of their molecular ions (Fig. 5.14). The corresponding individual pattern of each level of chlorination reflects perfectly the technical mixtures and can act for pattern recognition of technical products in natural samples. The same accounts for linear alkylbenzenes LABs (residues of detergents, frequently detected in riverine sediments), that are commonly analyzed and quantified using their characteristic ion m/z 91, where the individual homologues are verified by their molecular ion (Fig. 5.14).

5.1.3 Common Routine in GC/MS Based Identification

In common laboratory routine, GC/MS data are certainly not handled by ab initio interpretation of mass spectra or similar time consuming and intensive approaches. For identification of substances, IT based approaches are applied using mass spectral data libraries. These libraries consist of hundreds of thousands of mass spectra and computer-based comparison allows a fast search for similar spectra as compared to the measured ones. Precondition is the conformity of mass spectra measured on different mass spectrometer.

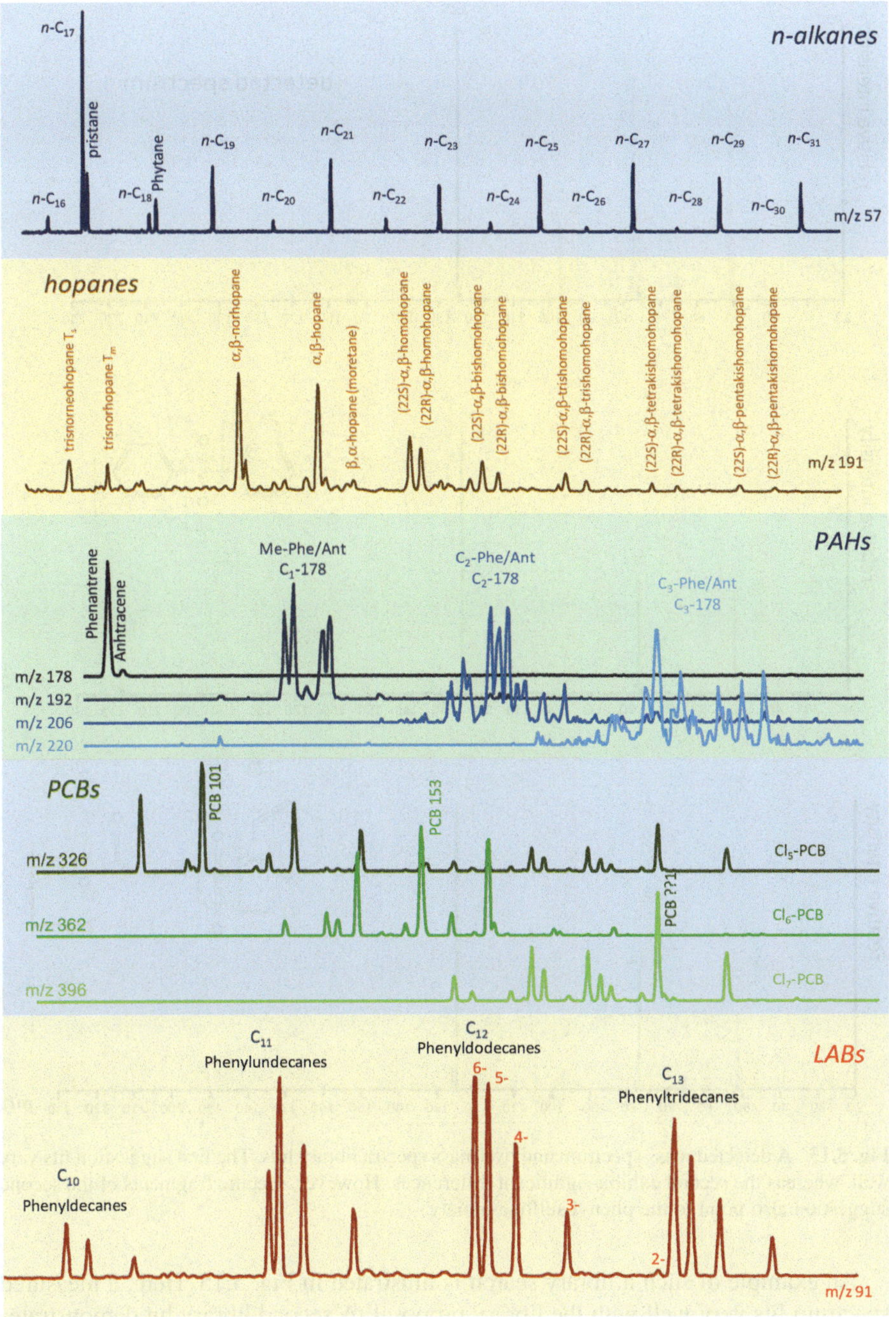

Fig. 5.14 Examples for ion chromatogram-based analyses in Organic Geochemistry: *n*-alkanes (**a**), hopanes (**b**), PAHs (**c**), PCBs (**d**) and LABs (**e**)

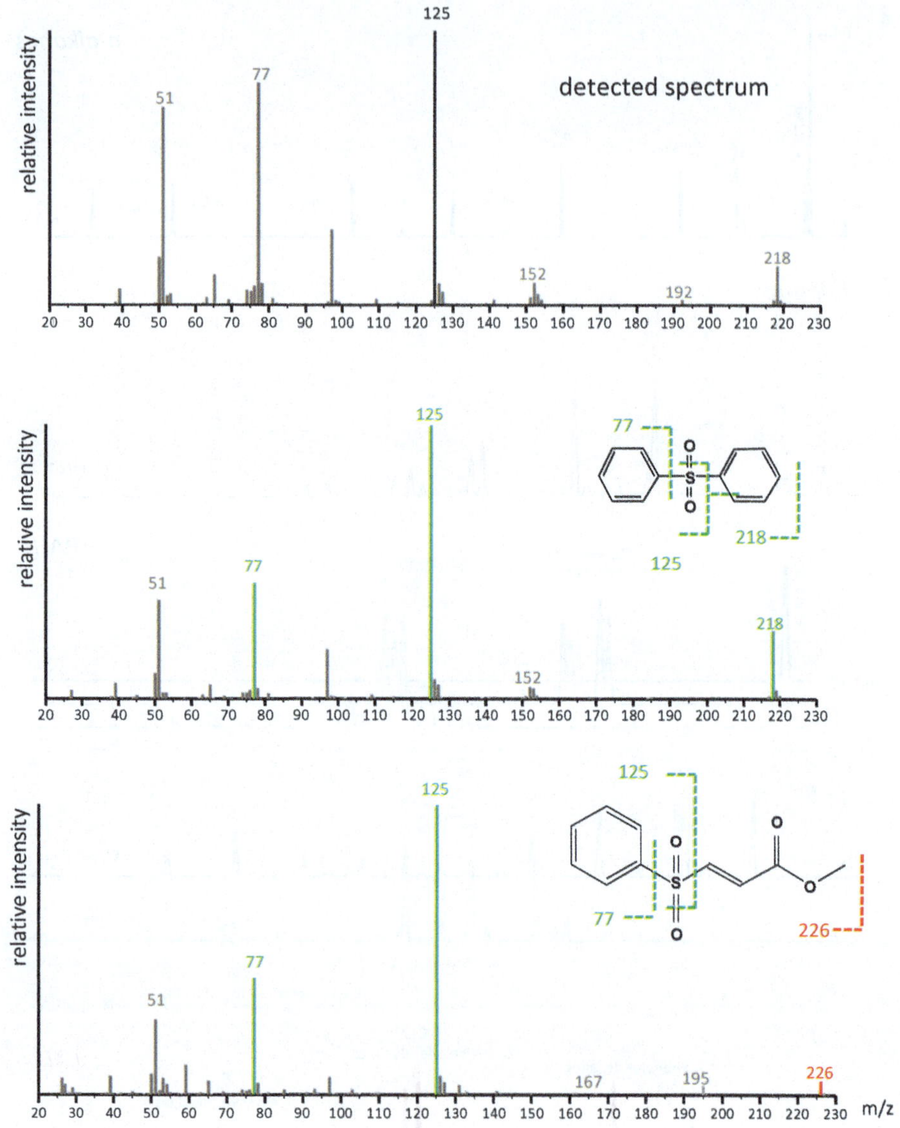

Fig. 5.15 A detected mass spectrum and two mass spectra library hits. The first suggestion fits very well, whereas the second exhibit significant differences. However, specific fragments of the second suggestion also point to the phenyl sulfone moiety

An example of such a library search is illustrated in Fig. 5.15. Here, a measured spectrum fits very well with the library proposal. A second library hit demonstrates also some similarity but not a perfect fit. However, the concurrent appearance of significant ions (here ions ...) point to similar structural moieties (here the phenysulfonic group). This exemplifies that also not perfectly fitting library suggestions can be used for obtaining partial structure information on the measured molecule.

The comparison with mass spectral libraries has certainly its limitations. High similarity does not imply clear evidence for identification. There are various examples pointing to be cautious with such simple interpretations. One example is related to positional isomers. Mass spectra of positional isomers (e.g. substitution at phenyl moieties) are often highly similar or even identical. An example is given in Fig. 5.16a. Here, the 4-methyl and 6-methyl isomers cannot be distinguished solely based on their mass spectra. Also the mass spectra of *n*-alkenes and corresponding *n*-alkanols are highly similar (see Fig. 5.16b) due to elimination of H_2O from alcohols in the ion source forming n-alkenes. Therefore, both substances exhibit nearly the same mass spectra and cannot be differentiated based on their mass spectral properties.

These examples point to more comprehensive requirements for an unambiguous identification of substances based on GC/MS. Noteworthy, for both examples a differentiation of the potential target compounds can be achieved by their chromatographic behavior. Positional isomers have slightly but different gas chromatographic retention times. Alkenes and alcohols exhibit not only different GC retention times but can be also easily separated by fractionation.

Therefore, an unambiguous identification of a compound by GC/MS can only be achieved by comparison of both mass spectral as well as gas chromatographic properties with those of a reference substance.

General Note
An organic substance is unambiguously identified by GC/MS if the mass spectrum and the gas chromatographic retention time match those of a reference substance under identical analytical conditions.

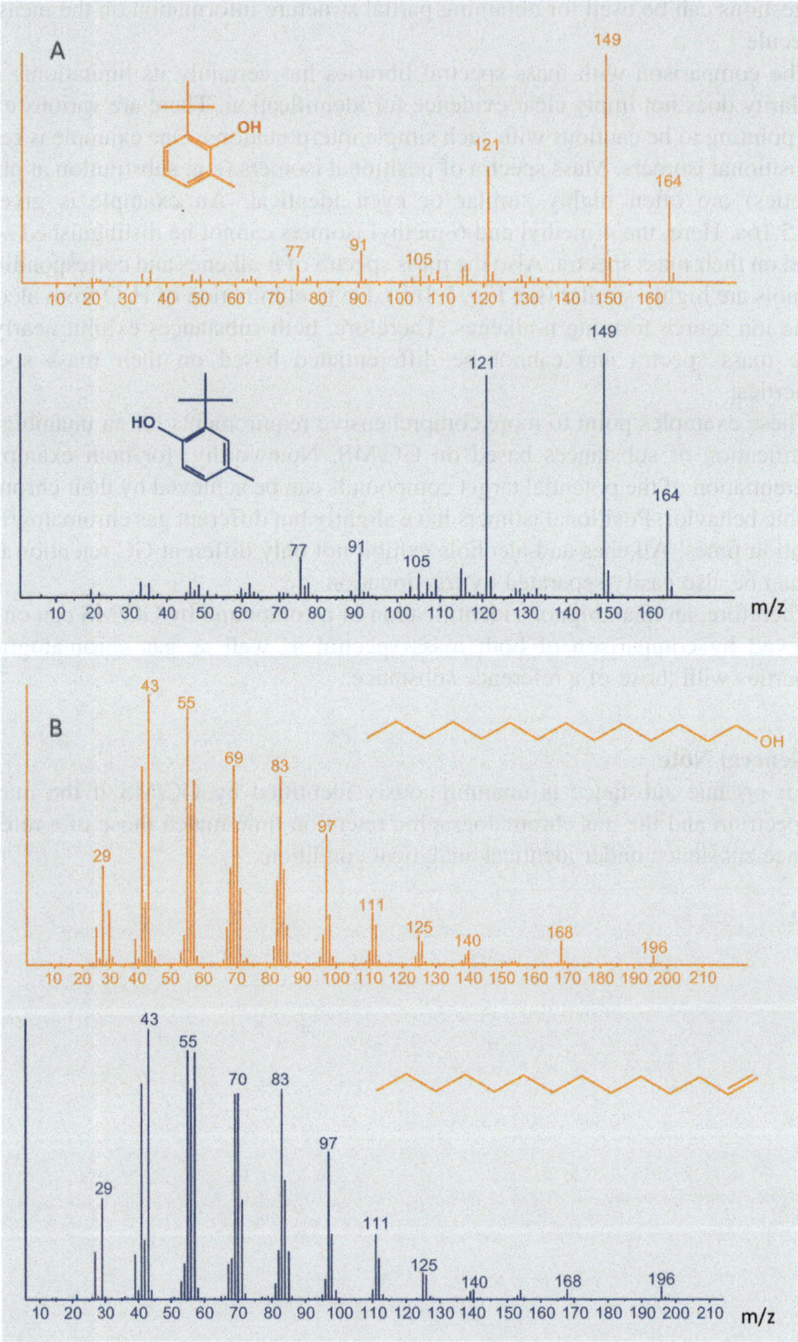

Fig. 5.16 Mass spectra of 6-methyl-2-isopropylphenol and 4-methyl-2-isopropylphenol (**a**) as well as 1-tetradecanol and 1-tetradecene (**b**) exhibiting high similarity, respectively

Excursus: ***What to do in case of missing reference material?***
If reference material is not available, secondary chemical information need to be considered for improving the evidence of identification. A principal scheme is given in the following:

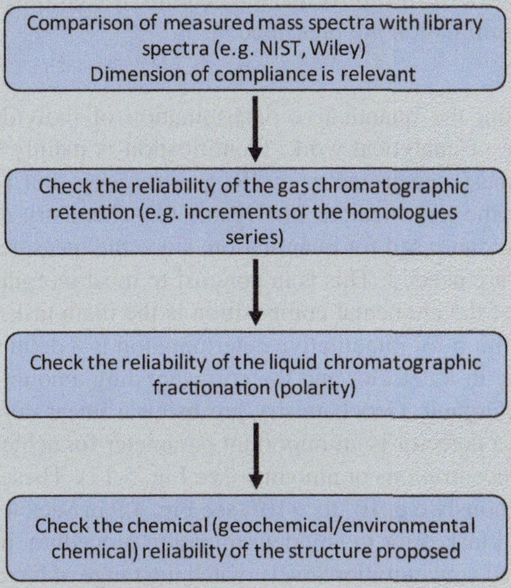

Please note, this approach gives higher evidence for identification but represents not an unambiguous identification!

As already mentioned, soft ionization as commonly established in particular in LC/MS systems does not induce fragmentation. Hence, identification based on fragmentation interpretation fails for these techniques. Dentification is based in these systems either strongly by verification with reference substances or by using high resolution mass spectrometry. The latter ones allow to determine elemental compositions but does not provide information on the complex structures. Generally, identification by LC/MS is so far not as powerful as compared to GC/MS based identification.

5.2 Quantitation

Outlook

In addition to identification, the process of quantitation is a main task of analytical work. Here, the principal procedures for chromatographic based quantitative determination are presented.

Beside identification, the quantitative determination of individual analytes is the second major issue of analytical work. Identification is mainly based on detection methods such as mass spectrometry or IR spectroscopy, but for quantitation the chromatography is the key method. In Organic Geochemistry dominantly GC/MS and LC/MS analyses are used for quantitation, since the measurement of individual organic molecules are needed. This is in contrast to most inorganic analyses, where the determination of the elemental composition is the main task.

A basic precondition for quantitative determination is a defined relation between the measured signal in a detector and the corresponding amount of analyte. Nearly all quantitation in Organic Geochemistry are using a linear correlation, hence the area of linearity of a detector is an important parameter for achieving a wider range of determinable concentrations or amounts (see Fig. 5.17). These ranges of linearity vary between enormously e.g. 10^3 to $>10^6$ (see Fig. 4.9 in Sect. 4.1). However, as an important analytical task prior to any determination procedure, it has to be checked whether the expected concentration levels match the range of linearity of the selected detector system.

However, a single signal (e.g. the sum of ions of a mass spectrum) is not sufficient for quantitation. The clue of quantitative determination in organic analyses is related to the chromatographic peaks. Chromatographic peaks represent a time resolved measurement of signals. As an important consequence, also the peaks (equivalent to individual signals) exhibit the same correlation to a measured concentration or

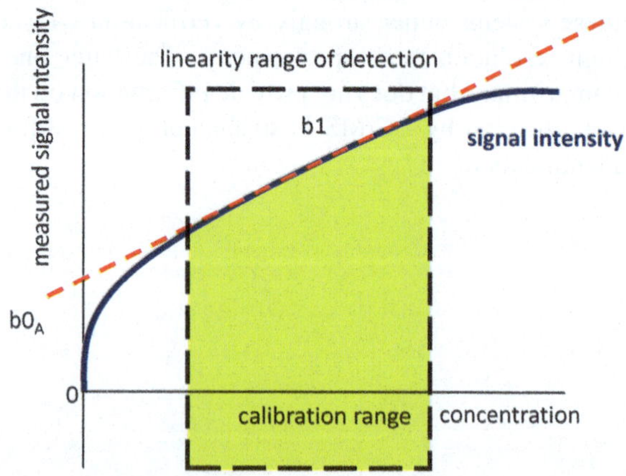

Fig. 5.17 The linearity range of a detector which represents also the calibration range

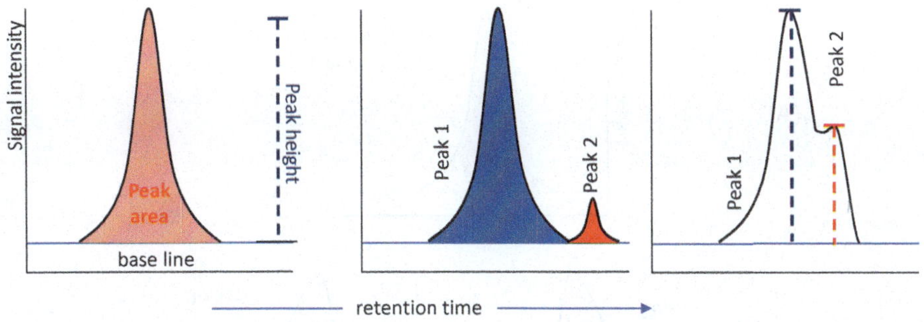

Fig. 5.18 Peak height and area as key parameter for quantification and two exemplary applications

amount, normally a linear correlation. This relation allows to link detected signals but also peaks and quantitative data. In chromatography (GC as well as LC) two main features of a peak can be used for quantitation, the peak height and the peak area (as illustrated in Fig. 5.18). The peak area is dominantly used for quantitation, but in some more special cases the peak height is favored. As an example, in case of a perfect chromatographic separation but also in case of asymmetric peaks the peak area can be easily calculated (Fig. 5.18b). But if the separation is insufficient leading to partly overlapped peaks (e.g. forming shoulders as illustrated in Fig. 5.18c), the peak height is a more precise parameter for quantitation.

This general approach is applied to GC and LC systems with more simple detectors (GC-FID or LC-fluorescence detector) measuring only one chromatogram but is preferably used with mass spectrometry as a multiple chromatogram detector (ion chromatograms). Peak areas can be obtained by integration, in nowadays done digitally. For a sufficient integration, the peak as primary quantitation tool needs to be of high quality. Perfect peaks exhibit a small peak width and especially a fully symmetric peak form (see also Sect. 3.3.1 and Fig. 3.5 as well as Fig. 5.19a). The two main effects lowering the peak symmetry are characterized as fronting or tailing (see Fig. 5.19a). Both effects increase the peak width which has an important negative implication for the peak integration (see Fig. 5.19b). Beside the peak width also the separation is an important aspect for peak integration. It is obvious that only a full chromatographic separation allows an exact integration of two neighbored peaks. Problems arise by a full or partial overlap of peaks as illustrated in Fig. 5.19c. Here only special detection approaches such as integrating different ion chromatograms for each analyte (see Sect. 5.1.2) or related SIR measurements (see Sect. 4.2.1) allow often a quantitation also of co-eluting analytes.

As already mentioned, nowadays peak area detection is performed by digital processes, however, in case of symmetric or slightly symmetric peaks the area can also be calculated by approximations using peak width at different peak heights and the height itself (as illustrated in Fig. 5.20).

One key issue of chromatography-based quantitation is to determine the correlation factor between detected peak area and amount or concentration. This process is called calibration and must be executed for each analytical system and each analyte.

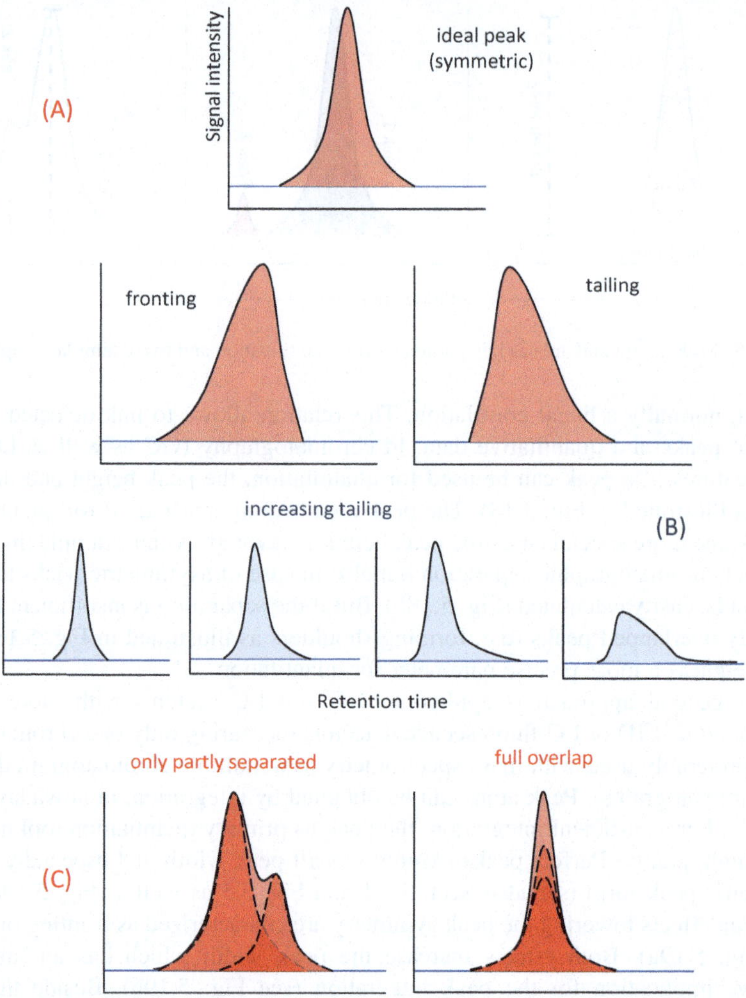

Fig. 5.19 Perfect peak form (**a**) and examples for negative shifts, (**c**) some pitfalls for peak-based quantification

It implies the calculation of the correlation factor, mostly the linear factor, that is represented by the gradient of the red line in Fig. 5.2. Consequently, a calibration is only valid in the linearity range of the detector.

> **General Note**
> Since the response factor is unique for an analytical system and an analyte, calibration needs to be performed for each system and each analyte individually.

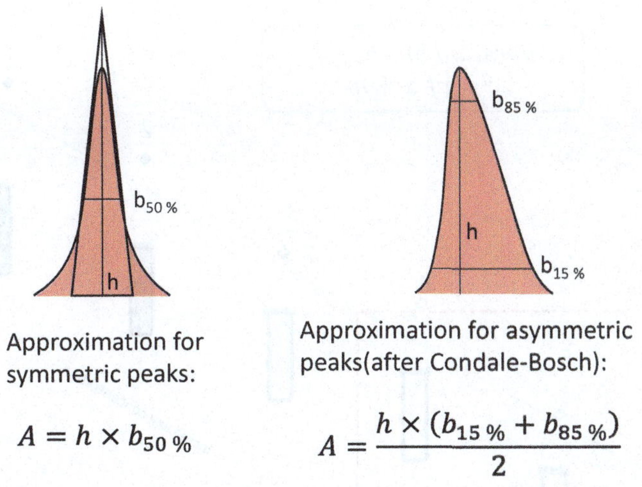

Fig. 5.20 Two approximations for calculating the areas of full symmetric or slightly asymmetric peaks

In principal three different calibration methods exist: an external and internal calibration as well as standard addition calibration. For an external calibration the measuring system gets calibrated by external standard solutions with different concentrations. The concentrations used should cover the expected unknown concentrations. Usually 4–5 calibration points are used, however the more the calibration points the better the calibration. With the resulting data set a linear correlation can provide a calibration function as illustrated in Fig. 5.21. The quality of the calibration is represented by the correlation coefficient, that should be near the value 1. Based on the calibration function, the concentration of the analyte in any further sample solution, normally extracts, can now be determined.

The second approach uses an internal standard for calibration. The internal standard should be a chemically highly similar substance such as labelled compounds (e.g. ^{13}C-labelled PCBs for PCB analyses) or unusual isomers or derivative not expected in the extract (e.g. theobromine for caffeine analysis). As an important prerequisite, these internal standard substances need to be detectable in parallel to the target analytes without any interference. A defined amount of internal standard is added to the sample solution or extract containing the target analyte with unknown concentration. After measurement, the peak area of the internal standard with known concentration is used as a one-point calibration to convert the peak area of the target analyte to its concentration. This process is exemplified in Fig. 5.22 for a PCB analyses based on GC/MS measurements.

The last calibration approach uses the analyte itself as standard but not as external but as added standard. This approach needs several analyses. First, the untreated extract is measured. Thereafter, a defined amount of analyte is added (usually by adding a defined volume of standard solution with known concentration) and the analysis is executed once again. This step of adding and measuring is repeated

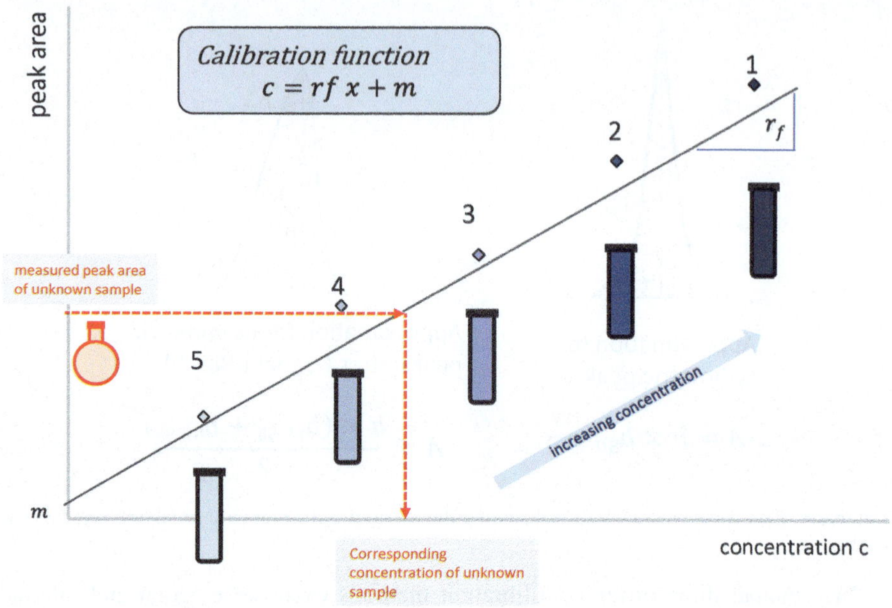

Fig. 5.21 Scheme of an external calibration procedure

several times (usually 4–5 times). Each step of standard addition creates a peak area difference that is correlated with the added concentration. Based on this correlation, a response factor can be calculated that can be applied to the peak area of the target analyte in the untreated extract (see Fig. 5.23).

Based on the three calibration methods, the concentration in an extract or sample can be determined but this value does not represent the concentration or amount in a natural sample (e.g. a river water or coal sample). To complete a quantitation, the extract concentration must be converted to the absolute amount of analyte by multiplication with the extract volume and, secondly, this resulting amount needs to be normalized to the total sample amount used for extraction. This procedure converts e.g. an extract concentration of several ng/µL to a water sample concentration of some ng/L or a soil contamination of some ng/g.

General Note
Quantitative analyses in Organic Geochemistry dominantly use chromatography. Quantitation follows on two steps, a peak integration and a calibration for converting the detected peak area to a concentration.

In Organic Geochemistry of fossil matter often biomarker analyses are performed. Here, a distinct determination of absolute quantitative data is not necessary, since ratios of chemically similar compounds are target parameters. For calculating such ratios, the peak areas can directly be used. However, two

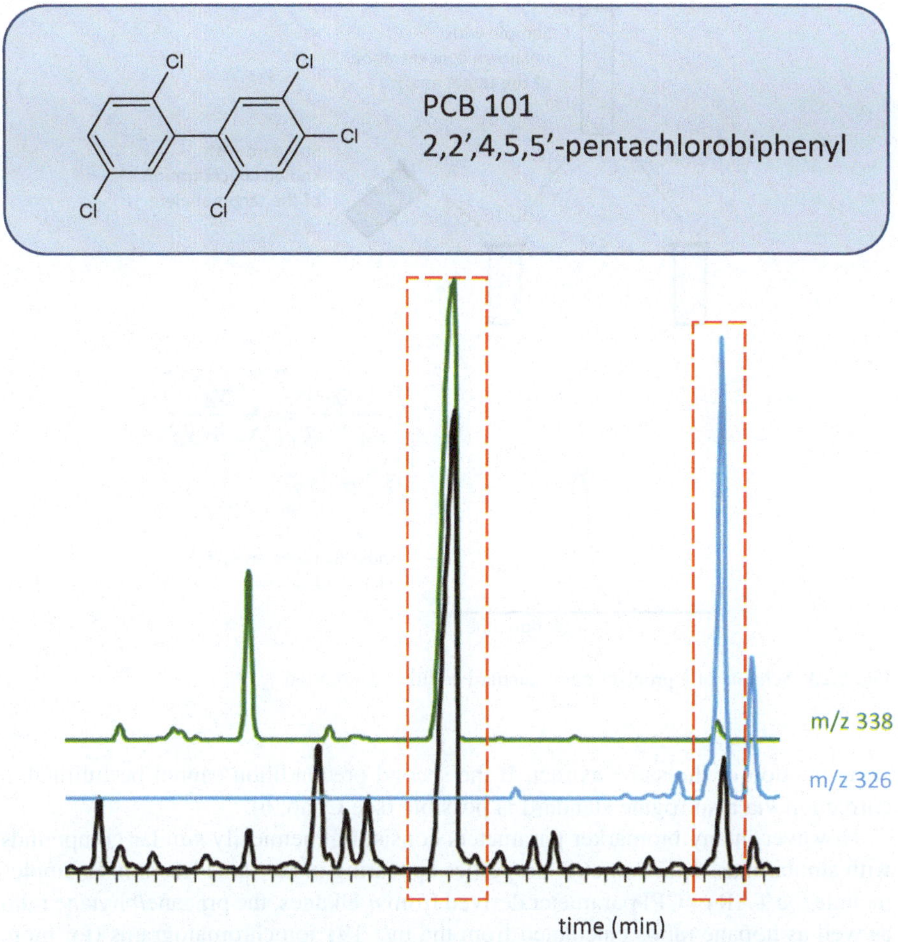

2,2′,4,5,5′-pentachlorobiphenyl (PCB 101)	analyte with unknown concentration
^{13}C labelled 2,2′,4,5,5′-pentachlorobiphenyl (+ m/z 12)	internal standard with known concentration

Fig. 5.22 Example of an internal calibration procedure for PCB analyses by GC/MS

preconditions are needed to be considered. Firstly, the response factors of the compounds have to be very similar or even identical. This accounts for chemically similar compounds and, in case of mass spectrometric analyses, similar mass spectra. Secondly, the compounds have to appear in the analyses, that implies to occur in the

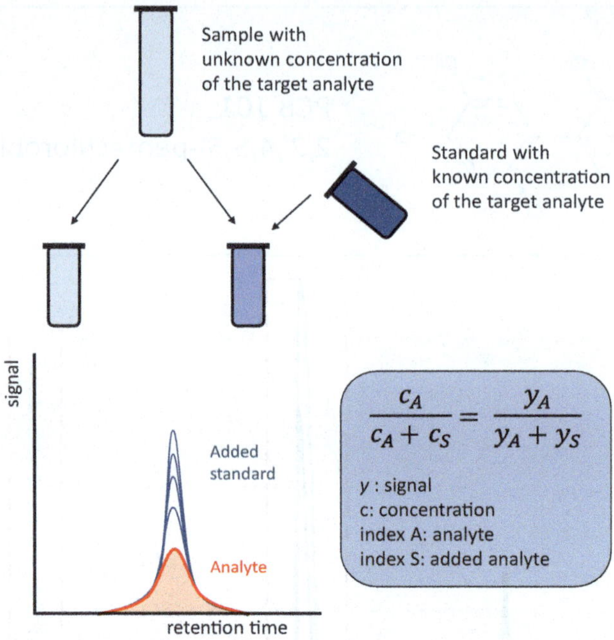

Fig. 5.23 Scheme of a procedure for standard addition calibration

same fraction or the same extract. If the second precondition cannot be fulfilled, a correction via a surrogate standard is possible (see Chap. 6).

However, many biomarker parameters consist of chemically similar compounds with similar mass spectrometric properties appearing in the same fraction. Examples include the TAR or CPI parameter derived from *n*-alkanes, the pristane/phytane ratio as well as hopane ratios calculated from the m/z 191 ion chromatograms (for more information on biomarker analyses see Volume 1 and 2 of *Fundamentals in Organic Geochemistry*).

Case Example: ***How to calculate a biomarker ratio***
Biomarker ratios are usually easily calculated. Base of the simple approach is the comparison of chemically highly similar substances exhibiting also very similar mass spectrometric properties. As an example, hopane ratios are commonly calculated from a specific ion chromatogram, the m/z 191 trace. An example is given in the following:

(continued)

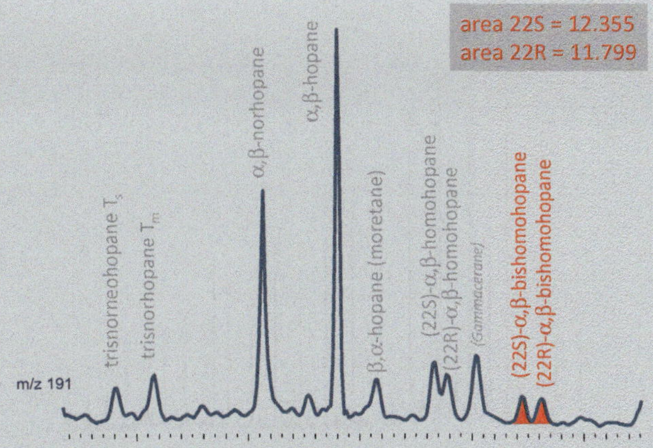

The maturity ratio 22S/22S+22R isomer of bishomohopanes can be determined by identifying the corresponding peaks in the m/z 191 ion chromatogram, by integrating the peaks to obtain the peak area and finally just by calculating the ratios by using simply the peak areas:

$$\frac{22S}{22S + 22R} = \frac{12.335}{12.335 + 11.799} = \mathbf{0.51}$$

Excursus: ***A very weak quantitation***

There exist one very scarcely used possibility for a rough estimation of quantitative amounts. Hereby, the peak area is related to the total area of a chromatogram, e.g. obtained by GC-FID analyses (see Figure below). The resulting value represents the rough estimation of the relative contribution of the target analyte to the overall amount in the extract. Noteworthy, due to the highly varying response factor of individual substances and considering the neglected proportion of the extract not detectable by the used analytical system (e.g. non-volatile substances in GC analyses), this method just gives an idea on the concentration level but no precise quantitative information.

It is partly used for detection of the so-called total petroleum hydrocarbons TPH obtainable by integrating the FID trace of the aliphatic fraction of an oil sample and roughly calibrated by an internal or external standard.

(continued)

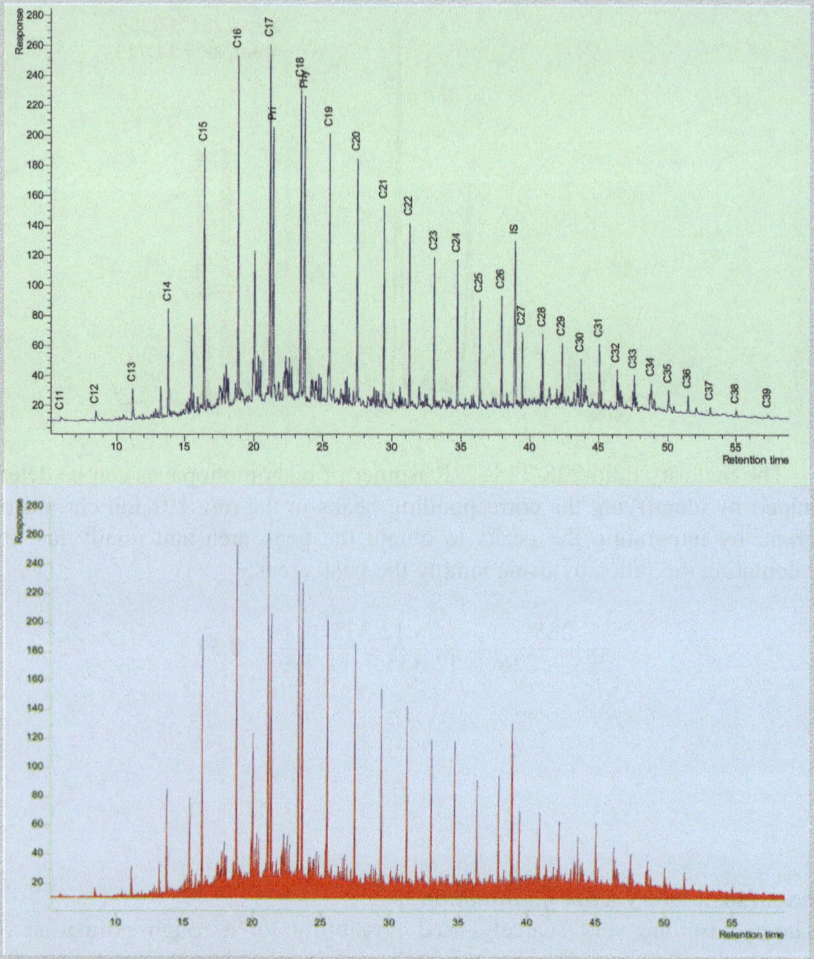

Figure: Example of a rough concentration estimation by integration of the total chromatogram without specific calibration

Chapter 6
Analytical Quality Control

Outlook
The final step of analysis is the evaluation of the obtained results by quality control. For quantitative data three parameters are important, the sensitivity, the accuracy and the reducibility. These aspects will be discussed in detail in this chapter.

All analytical data obtained need to be validated or evaluated. The quality control in qualitative analyses, that means the criteria for an unambiguous identification in Organic Geochemistry, have been already introduced in detail in Sect. 5.1. However, one aspect needs to be mentioned additionally. In order to avoid false positive detection as the result of laboratory contamination, the preparation and analyses of so-called blanks is recommended (see Fig. 6.2). These blanks are executed by using the whole laboratory working steps (extraction, evaporation, fractionation, filtration, . . .) but without any sample matrix. Analyses of such blanks always reveal substances normally at very low concentration levels that derive e.g. from the used materials (plasticizers, . . .), laboratory air (PCBs, solvents, . . .) or laboratory stuff (e.g. fragrances of perfumes, UV protectors, . . .). Substances identified in the blanks should be compared with the results of the original samples. All compounds detected in the blanks should not be considered as identified in the real samples except for significant differences in concentrations (e.g. magnitudes higher signals). However, the latter case needs to be checked very carefully.

Following the quality parameter for quantitative analyses are discussed in more detail. Reliable quantitative data are assessed by three parameters, the sensitivity of the analyses as well as their accuracy and reproducibility. Accuracy and reproducibility are often combined to the parameter precision. The sensitivity of an analysis can be expressed by two different values, the limit of detection LOD and the limit of quantitation LOQ. It has to be considered, for which analytical steps sensitivity has

J. Schwarzbauer, B. Jovančićević, *Introduction to Analytical Methods in Organic Geochemistry*, Fundamentals in Organic Geochemistry,
https://doi.org/10.1007/978-3-030-38592-7_6

been proven. One can address sensitivity for a target analyte determined by a defined analytical system e.g. GC/MS. But sensitivity can be determined also more comprehensively by including also the sample matrix.

A simple approach for determination of sensitivity for an analytical system, the data obtained from calibration processes can be used. Having a look on the external standard calibration approach, the basic component is the linear calibration function. The intersection of the linear curve with the x-axis marks a point at which for given concentration no peak area can be determined any more. This situation characterizes the limit of quantitation LOQ for the analytical system. Consequently, during external calibration the LOQ for a given system can easily be calculated. However, since this method solely is based on measurements of standard solutions, the matrix effects on the sensitivity are not considered. A second approach can be applied also for the measurements of real samples. It uses the noise of a measurements and its correlation with the signal intensity, the so-called signal-to-noise ratio (S/N ratio). Two different thresholds can be defined by this approach. The limit of detection as the lowest concentration, at which an unambiguous qualitative detection is possible, is defined by a minimum signal-to-noise ratio of 3 (see Fig. 6.1). The limit of quantitation as the lowest concentration, which can be determined quantitatively

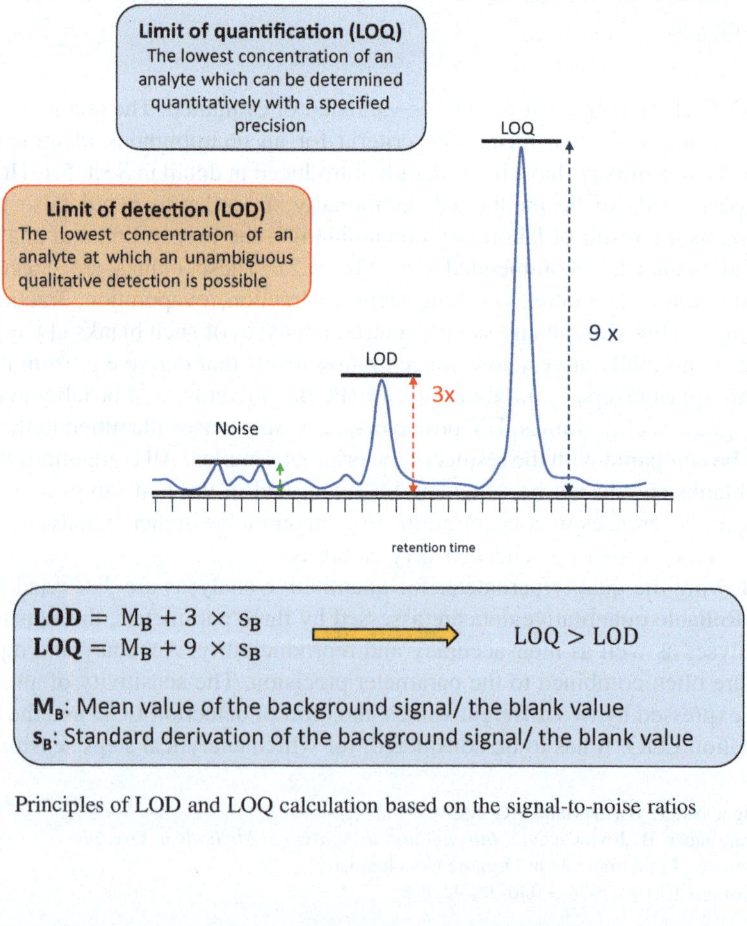

Fig. 6.1 Principles of LOD and LOQ calculation based on the signal-to-noise ratios

with a sufficient precision, needs a minimum signal-to-noise ratio of 9–10. To be accurate, the S/N ratios are calculated by the average values of the noise race and the addition of the corresponding standard deviations multiplied by factors of 3 for LOD and 9 (or sometimes 10) for LOQ. The S/N ratio approach can be applied to analyses of standard solutions revealing the sensitivity of the analytical systems (as also valid for the calibration curve approach). But measurements of real samples considering the matrix effects can also be used. This works sufficiently for samples with more or less homogenous matrices, but it is a huge challenge to calculate LOD or LOQ values for analyses of very heterogenous sample sets. Here, the matrix affects heavily and the noise level and, consequently, the calculated S/N ratios. In these cases, the LOQ can only be estimated by a thorough evaluation of all measurements.

To present data correctly, for all measurements with no signals or signals below the LOD the term 'not detected' or 'n.d.' should be used. If a signal falls between the both threshold values, the compound is identified or detected but cannot be quantified. Here, the usage of the term <LOQ is appropriate.

The quality parameter accuracy and reproducibility are often determined in one joint approach, the determination of recovery rates (see Fig. 6.2). For this purpose, precleaned or ultra-pure matrices (e.g. calcinated sea sand as substitute matrix for sediment or soil samples) are spiked with defined amounts of the target analytes. Then, the spiked samples are processed in the same way as the real samples and get analyzed. To calculate the recovery rate, the determined concentrations or amounts are related to the original spiked amounts or concentrations. These values reflect all losses during the analytical process. High recovery rates provide evidence for a sufficient accuracy of the method and can be used for correction of the values determined for the real samples. Such correction can also be achieved in a very smart way, if an internal standard calibration method is used. Adding the internal standard directly to the sample (water, soil, sediment, oil, . . .) prior to any analytical handling allows to determine indirectly all losses during the analytical procedure. Assuming the same loss for both target analyte and internal standard, the recovery rate is automatically considered by the calibration.

Another approach to correct losses and inaccuracies e.g. during evaporation or injection volumes, is the usage of surrogate standards (see Fig. 6.2). They are used similar to internal standards but are not applied for calibration. They are added to the original sample or individual fractions with known concentrations are traced through the analytical procedure till quantitative measurement. The discrepancy between initially added amount and finally determined one can also be used as correction factor to consider inaccuracies during the various process steps (extraction, fractionation, evaporation . . .).

Finally, the reproducibility or the variance of results can be evaluated by multiple recovery rate determination. Beside the average values (pointing to accuracy) the standard deviation reflects the reproducibility. Both parameters together characterize the precision as one key parameter assessing the quality of quantitative analyses.

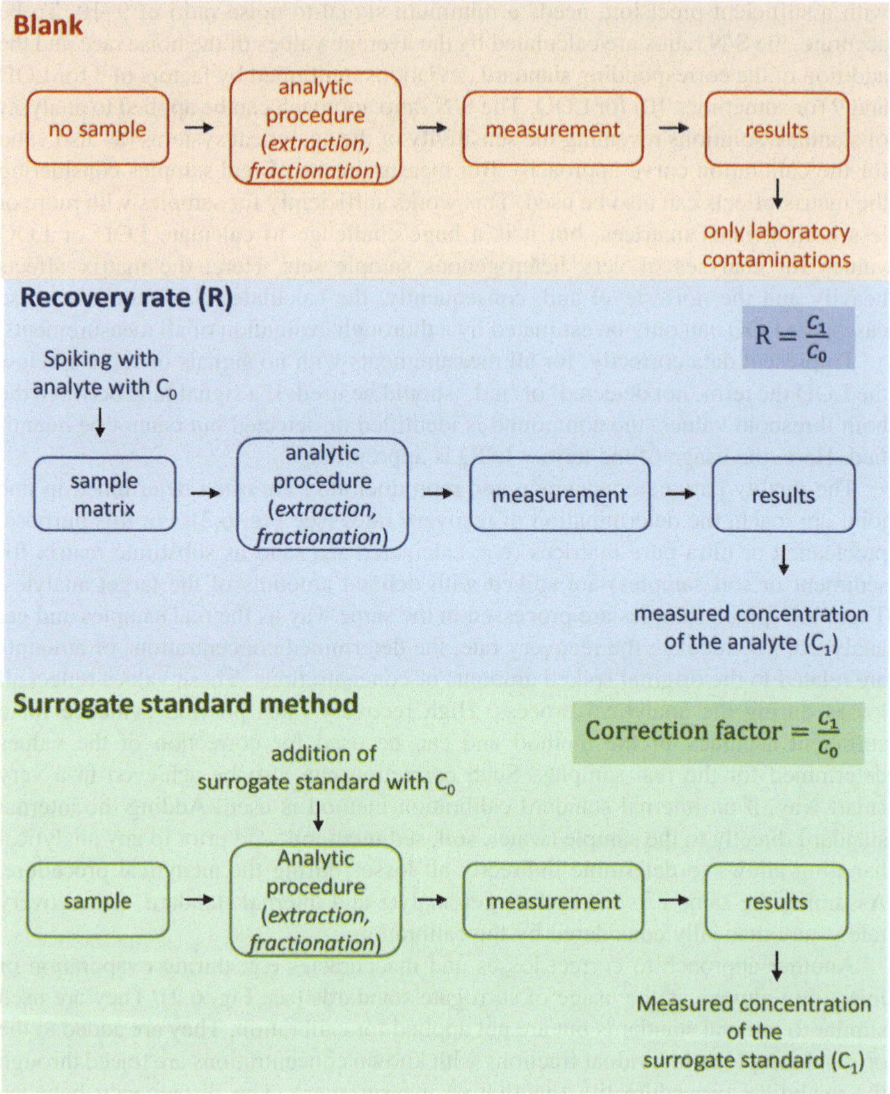

Fig. 6.2 Schemes of blank procedure, recovery rate and surrogate standard approaches

Case Example: ***A Comprehensive Quantitative Analysis***

As a final example, a quantification of an analyte X with quality assessment by a surrogate standard is described in the following.

An analysis applied to a 9.5 g soil sample included extraction, drying of the extract, fractionation, reduction of the fraction volume to 50 μL and

(continued)

subsequent measurement of a 1 μL aliquot by GC/MS. A surrogate standard has been added to the relevant fraction. The total procedure has been applied additionally to a precleaned sand sample spiked with the analyte at a concentration level that was expected also for the environmental sample. This recovery run has been performed three times revealing a recovery rate of 83% (±15% rel. standard deviation). A blank procedure performed on a precleaned sand sample revealed a laboratory contamination by the analyte at a level of 0.15 ng per analysis.

The GC/MS system has been calibrated with an external four-point calibration including the surrogate standard. The results from both, the calibration runs as well as the sample analyses are given in the following table:

Analysis	Concentration analyte (ng/μL)	Peak area (analyte)	Peak area (surrogate standard)	Peak area analyte normalized to surrogate standard
Calibration Std 1	50	1,250,970	100,050	12.50
Calibration Std 2	25	605,350	99,850	6.06
Calibration Std 3	10	255,580	101,150	2.53
Calibration Std 4	5	115,100	100,030	1.15
Soil extract	*Unknown*	550,780	98,950	5.57

For calibration, all areas of the analyte at the given concentrations were normalized to the surrogate standard area (just by division). These values are used for performing the linear correlation resulting in a calibration function:

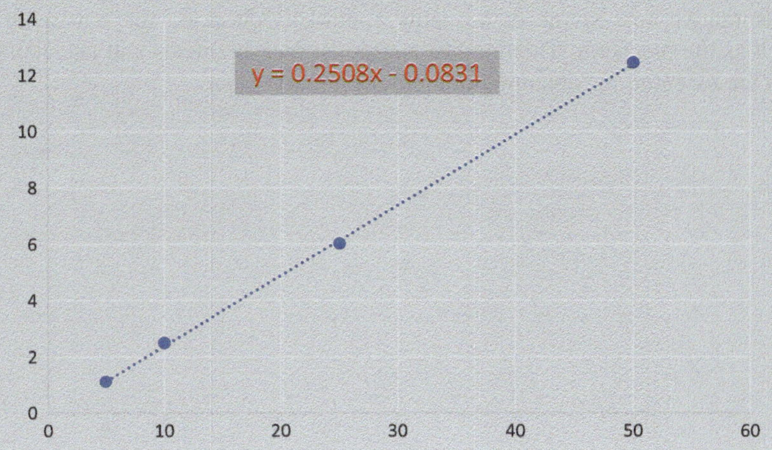

(continued)

The measured area of the analyte in the environmental sample is converted to absolute amount of analyte in the 1 μL aliquot by applying the calibration function. Thereafter, it is multiplied with 50 to obtain the total extracted amount of the analyte:

$$\text{Concentration of analyte in 1 μL} = \frac{5.57 + 0.0831}{0.2508} = 22.54\ \text{ng/μL}$$

$$\text{Total extracted amount of analyte} = 22.54\ \text{ng/μL} \cdot 50\ \text{μL} = 1122.5\ \text{ng} \approx 1120\ \text{ng}$$

A comparison with the blank value showed a neglectable contribution by laboratory contamination (at around 0.15 ng), hence the detected value is assumed to be reliable.

The total amount of the analyte already corrected by the surrogate standard is now normalized to the sample amount by division with 9.5 g (see above) to obtain the concentration in soil:

$$\text{Concentration of analyte in soil sample} = 1120\ \text{ng}/9.5\ \text{g soil} = 117.9\ \text{ng/g} \approx 118\ \text{ng/g}$$

Finally, this value is corrected by the recovery rate of approx. 83% resulting in the final value of 142.2 ng/g. Since the recovery experiments revealed a rel. deviation of around 10%, the result will be given as **140 ng/g**.

General Note

Quantitative analysis needs a quality control by checking the sensitivity as well as the precision. This implies additional measurements and calculations but are essential for obtaining reliable quantitative data.

Chapter 7
Principal Analytical Procedures in Organic Geochemistry

Outlook
In Organic Geochemistry a high variety of sample treatments and analytical techniques are used depending on the type of sample and the analytes focused on. However, some principal procedures are applied with variations in nearly all organic-geochemical laboratories that will be summarized in the following.

Generally, organic-geochemical laboratories are usually analyzing either fossil matter or environmental samples although the principal techniques used are comparable. This accounts in particular for the final detection methods, high-performance chromatography and coupled mass spectrometry (GC/MS and LC/MS). The corresponding strategies of identification and quantification are described in detail in Chap. 5. However, the type of analytes (e.g. pollutants, biomarker compounds . . .) and their varying concentration levels (approximately from μg/g to pg/g) require adapted sample preparation routines. Therefore, the corresponding analytical procedures are discussed here separately for both sample types excluding GC/MS or LC/MS detection. Noteworthy, also the available technical equipment as well as the individual experiences in different laboratories cause variations in analytical methods and procedures applied. The standard operation procedures given here as examples are derived from practical day-to-day routines in the laboratories of the authors.

J. Schwarzbauer, B. Jovančićević, *Introduction to Analytical Methods in Organic Geochemistry*, Fundamentals in Organic Geochemistry,
https://doi.org/10.1007/978-3-030-38592-7_7

7.1 Analyses of Fossil Matter

The procedures for oil analysis and for bitumen analysis from sedimentary rocks (including coal and petroleum pollutants in sediments or soil) are highly similar. The preparatory steps in bitumen analysis involves drying, grinding of the sample material and the bitumen isolation by an extraction method (Soxhlet extraction, ASE—see Sect. 3.2). Analysis of oil is preceded by the removal of water and suspended solid material as well as the precipitation and separation of so-called asphaltenes (very polar and macromolecular organic matter fraction). Following, the raw extracts or the pretreated oil sample are subjected to fractionation (see Sect. 3.3) and, finally, after reduction to a final volume the individual fractions get measured mainly by GC/MS but also by LC/MS. A scheme of these analytical procedures the preparatory operations are given in Fig. 7.1.

In the following some standard operation procedures for extraction and fractionation of fossil matter are listed. They derive from the authors' laboratories.

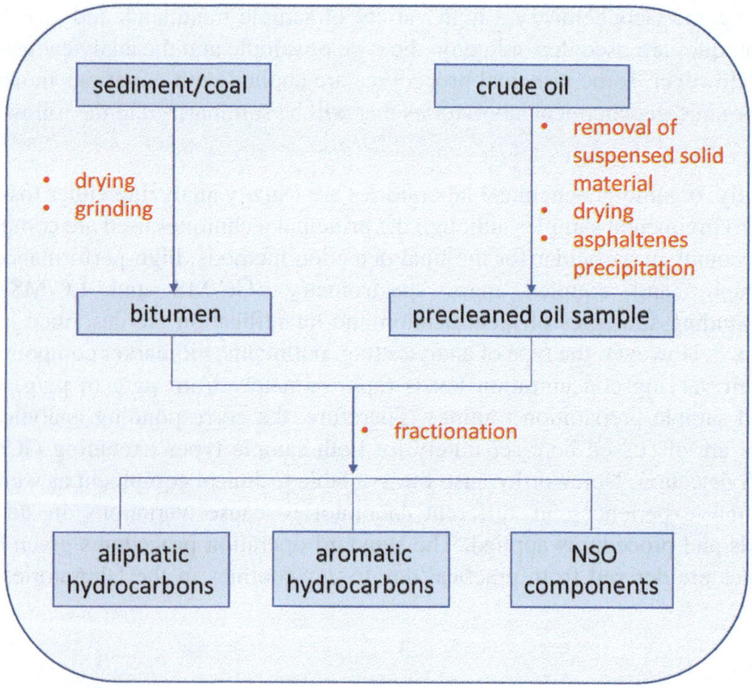

Fig. 7.1 Principal procedures for fossil matter analyses

Standard Operation Procedure SOP 1—Solid Matter Soxhlet Extraction (Rocks, Sediments, Soils, Coals . . .)

Preconditions:

- Clean all needed glass material by rinsing it with DCM and make sure to use only glass, metal and PTFE equipment.
- Use only precleaned (distillation) or ultra-pure solvents.
- Rinse the all parts of the Soxhlet device that will have contact with the sample.

Extraction:

- If applicable add an internal standard to the sample.
- Fill an appropriate amount of solid sample material (depending on the TOC value) into the extraction tube.
- Fill an appropriate amount of extractant (e.g. DCM, acetone . . . .) into the Soxhlet device.
- Start the first extraction cycle by heating the solvent.
- After 48 h of extraction the solvent can be changed to a less polar one (e.g. to *n*-hexane) if a more exhaustive extraction is needed. Start a second extraction cycle.
- After exhaustive extraction filtrate the raw extract and reduce the extract volume by rotary evaporation.
- Prepare a Pasteur pipette by half-filling with Na_2SO_4 (use glass wool as stopper at top and bottom, see on the right).
- After concentration to approx. 1 mL dry the raw extract over a Na_2SO_4 filled pipette and collect the dried extract.

- For desulphurization, add freshly activated copper powder (activation with HCl) to the sample and leave it in the refrigerator overnight.

Further handling:

Proceed with fractionation of the desulphurized raw extract

Standard Operation Procedure SOP 2—Solid Matter Extraction by ASE (Accelerated Solvent Extraction) (Rocks, Sediments, Soils, Coals . . .)
Preconditions:

- Clean all needed glass material by rinsing it with DCM and make sure to use only glass, metal and PTFE equipment.
- Use only precleaned (distillation) or ultra-pure solvents.
- Rinse the all parts of the ASE that will have contact with the sample.

Extraction:

- Choose an appropriate cell depending on the TOC content.
- If applicable add an internal standard to the sample.
- Fill the sample as well as beads into the precleaned cell. Make sure that the cell is filled completely.
- Start the automatic extraction cycle. Observe the pressure. In case of excessive pressure check if the filter is blocked.
- After extraction let the cell cool down, disassemble the cell completely and clean the filters and metal rings in the ultrasonic bath with organic solvent for 15 min.
- Concentrate the extract to approx. 1 mL.
- Switch extract solvents from DCM to *n*-hexane by adding approx. 2 mL of *n*-hexane to the raw extract and concentrate the sample by rotary evaporation. Repeat this procedure twice.
- Prepare a Pasteur pipette by half-filling with Na_2SO_4 (use glass wool as stopper at top and bottom)
- After concentration to approx. 1 mL dry the *n*-hexane phase over a Na_2SO_4 filled pipette and collect the dried extract
- For desulphurization, add freshly activated copper powder (activation with HCl) to the sample and leave it in the refrigerator overnight.

Further handling:
Proceed with fractionation of the desulphurized raw extract

Standard Operation Procedure SOP 3—Solid Matter Extraction by Ultrasonication (Rocks, Sediments, Soils, Coals . . .)
Preconditions:

- Clean all needed glass material by rinsing it with DCM and make sure to use only glass, metal and PTFE equipment.
- Use only precleaned (distillation) or ultra-pure solvents.

(continued)

Extraction:

- If applicable add an internal standard to the sample.
- Fill an appropriate amount of solid sample material (depending on the TOC value) into a round flask.
- Add an appropriate amount of a more polar extractant (e.g. DCM, acetone ….)
- Add a small piece of activated copper sheet for desulphurization.
- Add a magnetic stir bar to the flask and close the flask with a stopper.
- Treat the raw extract for 15 min in the ultrasonic bath.
- Afterwards use a magnetic stirrer and let it stir for 12 h.
- Use a pleated filter to separate the extract from the solids. Collect the extract separately.
- If an exhaustive extraction is required, repeat the extraction step with a second, less polar solvent.
- Concentrate the extract to approx. 1 mL.
- Switch extract solvents from DCM to *n*-hexane by adding approx. 2 mL of *n*-hexane to the raw extract and concentrate the sample by rotary evaporation. Repeat this procedure twice.
- Prepare a Pasteur pipette by half-filling with Na_2SO_4 (use glass wool as stopper at top and bottom)
- After concentration to approx. 1 mL dry the *n*-hexane phase over the Na_2SO_4 filled pipette and collect the dried extract.

Further handling:
Proceed with fractionation of the raw extract

Standard Operation Procedure SOP 4—Procedure for Fractionation of Fossil Matter Extracts

- Prepare a 2 g silica column. Use only high-quality silica gel. Activate the silica gel column in the oven at 200°C overnight.
- For conditioning add approx. 5 mL of n-pentane on the top of the column and let the solvent elute through the bottom of the column. Leave the column wet.
- Transfer the raw extract (reduced to a volume of approx. 0.5 mL) onto the top of the column. Let the sample penetrate completely the upper silica layers.
- Wash the empty extract vial with 0.5 mL of *n*-pentane and add the washing solvent also onto the top of the column.
- Add 5 mL n-pentane on the top of the column and collect the percolating solvent at the bottom. This is the first fraction (aliphatic hydrocarbons).

(continued)

- Repeat the elution with 5 mL of *n*-pentane/DCM (40:60 v:v) to collect the second fraction (aromatic hydrocarbons).
- Finally, repeat the elution with 5 mL of methanol to collect the third fraction (NSO components).
- Add an appropriate amount of surrogate standard to each fraction.
- Reduce the individual fractions to an appropriate analysis volume.

Further handling:
Measure all individual extracts separately by GC/MS or LC/MS

For special requirements or specific research questions some specialized procedure exist. As an example, the separation of nonbranched *n*-alkanes from branched isoprenoic aliphatics is required e.g. for successful compound specific stable isotope analyses, for increasing the sensitivity for isoprenoid analyses or for a successful analysis of *n*-alkanes in highly biodegraded oils. This separation is most frequently performed either with molecular sieve (5 Å) or with urea (so-called carbamide method or urea adduction). In both cases, n-alkanes are isolated by inclusion into pores or crystals with a diameter of 5 Å along with only *iso*- and *anteiso*-alkanes of long series. Because of their size, the molecules of other alkane fraction compounds (branched and cyclic alkanes) are not able to produce inclusion compounds neither with carbamide nor with these molecular sieves and remain separated in the extract. The result is visualized in Fig. 7.2.

Standard Operation Procedure SOP 5—Urea Adduction

- Evaporate the aliphatic fraction derived from the fractionation procedure to dryness.
- Redissolve the aliphatic fraction with a mixture of *n*-hexane and acetone (2:1 v:v)
- Add dropwise this solution to a saturated methanolic solution of urea.
- Separate the precipitate containing the urea adducts.
- Extract the nonadducts from the urea solution by *n*-hexane.
- Repeat the urea adduction with the nonadduct extract three times.
- Isolate the nonadducts in the residual *n*-hexane extract.
- Deliberate the adducts from the precipitated urea by dissolving it in water and subsequent extraction with *n*-hexane.

Further handling:
Measure both adducts and nonadducts separately by GC/MS

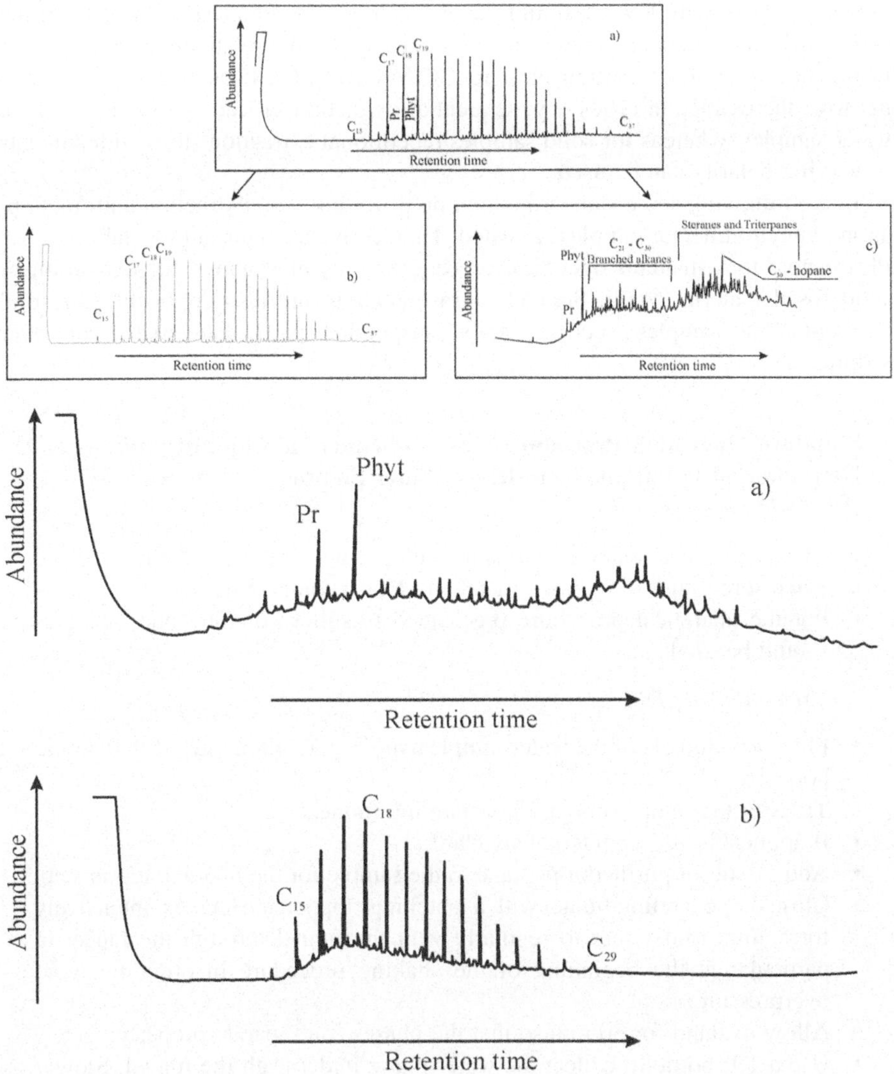

Fig. 7.2 Upper part—gas chromatograms of total alkane fraction of an oil (**a**), of the urea adducts (*n*-alkanes) (**b**) and of the urea non-adducts (branched and cyclic alkanes) (**c**). Lower part: Gas chromatograms of total alkane fraction of a biodegraded oil (**a**) and its urea adduct (*n*-alkanes) (**b**)

7.2 Analyses of Environmental Samples

Analysis in Environmental Organic Geochemistry dominantly focus on solid and liquid samples (mainly soils, sediments, dust, sludge as well as ground, surface and sewage water). For both sample types different preparation, extraction and fractionation procedures are needed as described in Chap. 3. Principally, water samples, especially those with low centration levels such as ground and river water, do not need a fractionation step, whereas many solid samples are relatively rich in organic matter needing a fractionation prior to GC/MS and LC/MS analyses. In order to improve the extraction efficiency, sequential extraction procedures are applied for water samples, whereas for solid samples repetitional extraction steps with solvents of varying polarity can be used.

In the following some standard operation procedures for extraction and fractionation of environmental samples are listed. They derive from the authors laboratories. Please note that standard operation procedures presented above for extraction of solid fossil material (ASE, Soxhlet, ultrasonication) are also applicable for solid environmental samples such as soils, suspended particulate matter or river sediments.

Standard Operation Procedure SOP 6—Sequential Liquid/Liquid Extraction of Pollutants from River Water Samples

Preconditions:

- Clean all needed material by rinsing with purified acetone and *n*-hexane and make sure to use only glass, metal and PTFE equipment.
- For the analytical procedure precleaned (distilled) or ultra-pure solvents should be used.

Extraction step1:

- Filter 1 L aliquots of the water sample using a glass fibre filter with 0.45 μm pore size.
- Transfer the sample into a 1 L separating funnel.
- If applicable add an internal standard.
- Add 50 mL of purified *n*-pentane to the sample for the first extraction step.
- Close the separating funnel with a matching stopper and shake it intensively for 5 min. Make sure to regularly vent the funnel through the faucet in particular at the beginning of the shaking procedure in order to avoid overpressure.
- Allow to stand for 30 min so that the phases can separate properly.
- Use a 1 L bottle to collect the water phase underneath the funnel. Slowly open the valve and close it right before the *n*-pentane phase enters the bottle.
- Decant the *n*-pentane phase through the top opening of the funnel into a small flask.
- Open the valve of the funnel and dry it with approximately 1 pipette of acetone.
- If applicable add a surrogate standard to the *n*-pentane phase.
- Concentrate the organic extract by rotary evaporation.

(continued)

- Prepare a Pasteur pipette by half-filling with Na_2SO_4 (use glass wool as stopper at top and bottom)
- After concentration to approx. 1 mL dry the *n*-pentane phase over the Na_2SO_4 filled pipette and collect the dried extract.
- Rinse the flask at least three times with *n*-pentane including drying over Na_2SO_4. Collect the rinsed *n*-pentane and combine with the extract.
- Concentrate the combined *n*-pentane extract to a final analysis volume of approx. 50 µL for further GC and GC/MS measurements.

Extraction step 2:

- Fill the pre-extracted water sample again into the separation funnel and add 50 mL of DCM.
- Shake the funnel for 5 min and vent it sporadically.
- Use a new flask to collect the DCM phase.
- Open the valve of the separating funnel and slowly separate the DCM Phase. Close the valve before the water phase can enter the DCM phase in the flask.
- If applicable add a surrogate internal standard.
- Dry and concentrate the DCM extract analogously to the pentane extract.

Extraction step 3:

- Lower the pH-value of the pre-extracted water sample to 2 by adding HCl.
- Treated the acidified water sample in the same way as the second extract.

Further handling:

Measure all individual extracts (extract 1: *n*-pentane; extract 2: DCM; extract 3: DCM at pH2) separately by GC/MS or LC/MS

Standard Operation Procedure SOP 7—Solid Phase Extraction (SPE) of Pollutants from Water Samples (River Water, Ground Water, Sewage Effluents . . .)

Preconditions:

- Clean all needed material by rinsing with purified acetone and *n*-hexane and make sure to use only glass, metal and PTFE equipment.
- For the analytical procedure precleaned (distilled) or ultra-pure solvents should be used.

(continued)

Extraction:

- Use an aliquot of 150 mL of the water sample.
- Prepare a SPE column with appropriate amount of adsorbent or take a commercially available SPE cartridge.
- Condition the column according to the instructions given by the SPE supplier.
- Connect the column to a vacuum pump and hold the column upside down into the water sample. The flow rate should not exceed 3 mL/min.
- Pump the whole sample volume through the cartridge or column.
- Thereafter, dry the cartridge or column by leaving it connected to the vacuum pump and allow to pump air for approx. 5 min.
- Elute 5 mL of DCM through the column and collect the extract.
- If applicable add a surrogate standard.
- Prepare a Pasteur pipette by half-filling with Na_2SO_4 (use glass wool as stopper at top and bottom)
- After concentration to approx. 1 mL dry the extract over the Na_2SO_4 filled pipette and collect the dried extract.

Measurements:
Measure the extract by GC/MS or LC/MS

Standard Operation Procedure SOP 8—Fractionation of Environmental Solid Matter Extracts (Soils, Sediments . . .)

- Prepare a 2 g silica column. Use only high-quality silica gel. Activate the silica gel column in the oven at 200°C overnight.
- For conditioning add approx. 5 mL of n-pentane on the top of the column and let the solvent elute through the bottom of the column. Leave the column wet.
- Transfer the raw extract (reduced to a volume of approx. 0.5 mL) onto the top of the column. Let the sample penetrate completely the upper silica layers.
- Wash the empty extract vial with 0.5 mL of *n*-pentane and add the washing solvent also onto the top of the column.
- Add 5 mL n-pentane on the top of the column and collect the percolating solvent at the bottom as the first fraction.
- Repeat the elution with 8.5 mL of *n*-pentane/DCM (95:5 v:v) to collect the second fraction.
- Repeat the elution with 5 mL of *n*-pentane/DCM (90:10 v:v) to collect the third fraction.

(continued)

- Repeat the elution with 5 mL of *n*-pentane/DCM (40:60 v:v) to collect the fourth fraction.
- Repeat the elution with 5 mL of DCM to collect the fifth fraction.
- Finally, repeat the elution with 5 mL of methanol to collect the sixth fraction.
- Add an appropriate amount of surrogate standard to each fraction.
- Reduce the individual fractions to an appropriate analysis volume.

Further handling:
Measure all individual extracts separately by GC/MS or LC/MS

The manufacturer's authorised representative in the EU is Springer Nature Customer Service Centre GmbH, Europaplatz 3, 69115 Heidelberg, Germany. If you have any concerns regarding our products, please contact ProductSafety@springernature.com

Printed and bound by CPI Group (UK) Ltd, Croydon, CR0 4YY
30/09/2026
02225465-0009